Magnesium

From Resources to Production

Magnesium

From Resources to Production

Hussein K. Abdel-Aal

CRC Press
Taylor & Francis Group
Boca Raton London New York

CRC Press is an imprint of the
Taylor & Francis Group, an **informa** business

CRC Press
Taylor & Francis Group
6000 Broken Sound Parkway NW, Suite 300
Boca Raton, FL 33487-2742

First issued in paperback 2020

© 2019 by Taylor & Francis Group, LLC
CRC Press is an imprint of Taylor & Francis Group, an Informa business

No claim to original U.S. Government works

ISBN 13: 978-0-367-65719-2 (pbk)
ISBN 13: 978-0-8153-4633-3 (hbk)

Library of Congress Cataloging-in-Publication Data

Names: Abdel-Aal, Hussein K., author.
Title: Magnesium : from resources to production / Hussein K. Abdel-Aal.
Description: First edition. | Boca Raton, FL : CRC Press/Taylor & Francis
Group, 2018. | "A CRC title, part of the Taylor & Francis imprint, a
member of the Taylor & Francis Group, the academic division of T&F Informa
plc." | Includes bibliographical references.
Identifiers: LCCN 2018006574 | ISBN 9780815346333 (hardback : acid-free paper)
Subjects: LCSH: Magnesium.
Classification: LCC TA480.M3 A23 2018 | DDC 661/.0392--dc23
LC record available at https://lccn.loc.gov/2018006574

Visit the Taylor & Francis Web site at
http://www.taylorandfrancis.com

and the CRC Press Web site at
http://www.crcpress.com

Contents

Preface

If I had to go back and reselect a title for my book; I would have given it the title: *Magnesium the Dream, or Magnesium: Metal of the Future.* Two reasons justify this title:

- Today, magnesium is bringing a new revolution in the energy domain as an alternative new source of energy, which is a kind of dream!

- It is a dream to see that this book has published. Four decades have elapsed since our first experiment was carried out at KFUPM laboratory to recover magnesium from brine by the Preferential Salt Separation (PSS).

- Magnesium is described as the "lamp of life. It captures the light energy from the sun. Magnesium is bound as the central atom of the porphyrin ring of the green plant pigment chlorophyll. It is the element that causes plants to be able to convert light into energy.

Magnesium is the metal of the future! This post was found on the web from an enthusiastic magnesium believer. It fairly sums up the consensus surrounding the future of magnesium. Magnesium is 33% lighter than *aluminum*, 60% lighter than *titanium*, and 75% lighter than *steel*.

Despite the fact that it is easy to find, magnesium is never found free in nature. As a result, many different methods have been developed to separate magnesium from other substances. Magnesium would be described as a *friendly or versatile element*, since it bonds with other elements. In other words, it does not live in isolation. Good examples are $MgCl_2$ and MgO. Because of its strong reactivity, it does not occur in its native state, but rather it is found in a wide variety of compounds in seawater, brines, and rocks. Its reactivity is the reason why this very plentiful element has not been produced as a metal earlier in civilized history, and is also responsible for certain erroneous ideas that tend to limit its commercial use today.

This friendship is what costs us money and energy to extract it and to obtain it as a *free* element.

Much research on magnesium processes is being performed in many areas throughout the world. The exploration into more efficient methods is bordering on a very large breakthrough to a very efficient and lower cost process. There are few practical examples of the further processing of brines from desalinization plants to obtain magnesium chloride. With the available technology we have today, this can change a *waste product* into a *resource*, aiming for the utilization of brine, as a feed stock for magnesium production, to take place.

Magnesium: From Resources to Production provides professionals, practicing engineers, and students of minerals processing, chemical engineering, and chemistry with a thorough and comprehensive knowledge on the different phases underlying the magnesium technology. The text should provide engineers with the know-how and tools to unearth and seize opportunities in this field.

The book is written in a manner that takes an in-depth look at the technologies, processes, and capabilities to suggest ways to overcome some of the limitations and challenges associated with the extraction process of magnesium. The most significant developments are therefore in extraction technology, where new processes should reduce extraction costs. It is hoped that the preferential salt separation (PSS) may take over to be the primary extraction process for the direct recovery of magnesium as magnesium chloride from seawater.

Extraction is the key process that makes magnesium available from its resources. It simply means to remove or take out, especially by effort or force. For the case of seawater, the extraction process of magnesium as end product consists of two consecutive steps:

1. Extraction of magnesium ions from seawater, by either chemical reactions (Dow Process) or by the physical separation (PSS) to have magnesium in the form of $MgCl_2$
2. Extraction of magnesium metal from $MgCl_2$ by the electrolytic process or the thermal process

The book consists of eight chapters. Chapter 1 is an introduction; it is a kind of bird's eye view of magnesium. Occurrence and resources of magnesium are described in Chapter 2. Chapter 3 represents an exploration study on magnesium and mineral salts from seawater. Chapters 4 and 5 are devoted to magnesium chloride, the backbone compound of the magnesium industry. Commercial methods for magnesium production as an end-product is detailed in Chapter 6. Chapter 7 represents a source of information on current applications of magnesium and prospective ones in the energy domain. Most importantly, the chapter captures the reader's interest on magnesium as a source of power for our global needs. Chapter 8 provides an account for the use of magnesium to store hydrogen in the form of magnesium hydride. Finally, "Facts and Uses" about magnesium is compiled in Appendix A.

Acknowledgments

All praise goes to *Almighty Allah*, who is the creator of this universe. I am grateful to God for the good health and well-being that were necessary to complete this book.

I would like to thank my wife for standing beside me throughout my career and for writing my fourth book. She has been my inspiration and motivation for continuing to move my career forward.

I like to thank my colleagues who got involved in the experimental work and provided insight and expertise that greatly assisted my research. Dr. Maha Abdel Kreem and Dr. Khaled Zohdi, both at Higher Technological Institute, Tenth of Ramadan City, Egypt.

I truly *acknowledge* the valuable time I had with the late Dr. M. Kettani, who was a close friend and a partner in doing our first experiment on the recovery of magnesium from seawater at KFUPM, Dhahran, Saudi Arabia.

I am grateful to Professor Mahmoud El-Halwagi at Texas A&M University and Professor Halim Redhwi at KFUPM, Saudi Arabia, for their support and encouragement in writing the proposal of the text.

A word of appreciation goes to Allison Shatkin and Teresita Munoz at Taylor & Francis Group for their cooperative support and dedication in getting this book in the hands of the reader.

Author

Dr. Hussein K. Abdel-Aal is Emeritus Professor of Chemical Engineering & Petroleum Refining, NRC, Cairo, Egypt. Prof. Abdel-Aal received his MS and PhD in Chemical Engineering in 1962 and 1965, respectively, from Texas A&M University, USA.

Professor. Abdel-Aal joined the department of Chemical Engineering at KFUPM, Dhahran, Saudi Arabia (1971–1985). He was the head of department for the period 1972–1974. Next, he was a visiting professor with the Chemical Engineering Department at Texas A&M (1980–1981). In 1985–1988, Prof. Abdel-Aal assumed the responsibilities of the head of Solar Energy department, NRC, Cairo, before rejoining KFUPM, for the period of 1988–1998.

Professor. Abdel-Aal conducted and coordinated projects involving a wide range of process development, feasibility studies, industrial research problems, and continuing-education programs for many organizations in the Middle East.

Prof. Abdel-Aal has contributed to over 90 technical papers and is the editor of *Petroleum Economics & Engineering, 3rd edition*, 2014, the main author of a textbook entitled *Petroleum and Gas Field Processing, 2nd edition*, 2016, and the sole author of *Chemical Engineering Primer with Computer Applications*, 2016; all three books are published by Taylor & Francis Group/CRC Press.

Professor Abdel-Aal is listed in *"Who is Who in the World,"* 1982, and is a member of AICHE, Sigma Si, and Phi Lambda Upsilon. He is a fellow and founding member of the board of directors of the International Association of Hydrogen Energy.

1

Introduction

1.1 Introduction

Land and fossil fuel resources are limited on our planet. Therefore, it is significantly important to harness the sun and sea for the creation of new resources. Magnesium recovery from seawater is one such area that represents a potential target. The oceans are vast and contain immense amounts of magnesium. The sea is considered a *store house* for magnesium since it contains a minimum of one million tons of magnesium per cubic kilometer. This means about 1.7×10^{24} tons of magnesium is stored in the oceans, which, in principle, could be recovered without the need of digging, crushing, processing, and all the other complex and energy-expensive procedures that we need for land mining.

Magnesium is described as having a wide range of uses in industry. In addition, it is an important element in the field of medicine. For industrial applications, it is often used with aluminum in the form of an alloy. Adding magnesium in the form of an alloy lightens the weight of aluminum while keeping its mechanical, fabrication, and welding characteristics intact.

Magnesium is well-ranked as:

- The *eighth* most abundant element in the earth's crust
- The *fourth* most common element in the earth, following iron, oxygen, and silicon
- The *eleventh* most abundant element by mass in the human body
- The *third* most abundant element dissolved in seawater, where sodium and chlorine are ranked first and second, respectively

Other additional features that magnesium enjoys are:

- About 13% of the planet's mass is made of magnesium.
- About 2.1% of the earth's crust is made up of magnesium, which is a sign of its abundance.

- Abundant in nature, mainly in seawater and brines as positive ions.
- Essential to all cells and some 300 enzymes.
- Abundant in the terrestrial crust in the magnesite form ($MgCO_3$) and dolomite ($MgCO_3 \cdot CaCO_3$).

1.2 Resources of Magnesium

Magnesium (Mg) exists in seawater as ions of magnesium. Most importantly, magnesium chloride is found in seawater, brines, and salt wells, as seen in Figure 1.1.

Sodium, chloride, magnesium, sulfate, and calcium are the most abundant dissolved ions in seawater. Magnesium is the third most abundant element dissolved in seawater after sodium and chlorine. The chemical composition of seawater, as given in Table 1.1, indicates the magnesium content in seawater.

Globally, resources for Mg could be described as *virtually* unlimited. Dolomite and magnesium-bearing evaporite-minerals are plentiful.

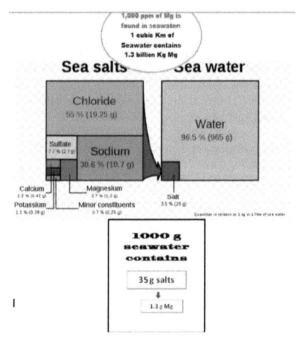

FIGURE 1.1
Magnesium concentration profile in seawater.

TABLE 1.1

Major Ion Composition of Seawater (mg/L)

	Typical Seawater	Eastern Mediterranean	Arabian Gulf at Kuwait	Red Sea at Jeddah
Chloride (Cl⁻)	18.980	21.200	23.000	22.219
Sodium (Na⁺)	10.556	11.800	15.850	14.255
Sulfate (SO₄²⁻)	2.649	2.950	3.200	3.078
Magnesium (Mg²⁺)	1.262	1.403	1.765	742
Calcium (Ca²⁺)	400	423	500	225

Source: https://www.soest.hawaii.edu/oceanography/courses/OCN623/Spring%202015/
Salinity2015web.pdf.

Magnesium is found as its oxide, being a constituent of rocks in different ore forms. Magnesium-bearing brines constitute a huge resource in the billions of tons. Finally, and most important, is the source of Mg from seawater. Magnesium could be described as a *friendly* element. This *friendship* exhibited by magnesium indicates the fact that magnesium exists combined with other elements. In other words, it associates with other elements. It does not occur uncombined in nature. It is not found in a pure form because it bonds with other elements.

1.3 How to Obtain Magnesium?

To obtain magnesium from its resources, we have to apply energy in order to retrieve a usable amount of Mg. In this context, it was noticed that the word *extraction* has been improperly used in the literature. *Extraction* is the key process that makes magnesium available from these resources. The word *extract* simply means to remove or take out, especially by effort or force. For the extraction of magnesium from its ores, a suitable process must be designed using the proper type of energy. When it comes to seawater, the chemical approach to obtain magnesium as magnesium chloride is the current option. Here, the energy of the formation of the participating chemicals is utilized in the separation process. Magnesium is extracted from seawater by precipitating it as magnesium hydroxide first. Next, hydrochloric acid is added. The acid reacts with the hydroxide, forming magnesium chloride as an end product. This process is well-known on the industrial scale as the *Dow process*.

The other option is to use the proposed Preferential Salt Separation (PSS), or the physical method (proposed by the author and coworkers) to separate magnesium directly as magnesium chloride. This approach utilizes thermal energy in evaporating seawater.

This extraction process represents only one stage in obtaining magnesium. The other important stage is to electrolyze the magnesium chloride by drying this product to 1.5 moles of water; that is, make it anhydrous and then introduce it directly into the electrolysis cells. Here, electrolysis energy is used in order to obtain magnesium as the end product:

$$MgCl_2 + Energy \rightarrow Mg + Cl_2$$

Basically, the two steps outlined earlier represent the amalgamated extraction process of obtaining magnesium, as the end product, from seawater, as a resource, as shown in Figure 1.2.

Figure 1.3, on the other hand, describes both methods (Chemical and Physical PSS) of extracting magnesium, as the end product, from seawater by the electrolytic process.

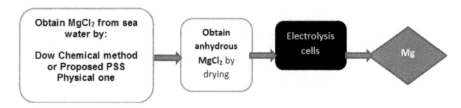

FIGURE 1.2
Main two steps leading to the production of magnesium metal.

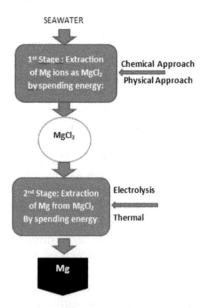

FIGURE 1.3
Extraction of magnesium metal from seawater.

Since the discovery of magnesium by Sir Humphry Davy in 1808, electrolytic magnesium production has been the major method used in the production of magnesium metal. Some developments by the early workers were accomplished. However, the electrolysis of anhydrous magnesium chloride proved to be the standard method of production. Since then, recognition of the process became widespread in application.

Most of the large magnesium production plants in the world today use electrolytic processes, which generally require magnesium chloride as a feed and use an electric current to convert this to magnesium metal and chlorine gas. Some of them use a feed source, such as brine or evaporated bitterns, to produce magnesium chloride.

On the other hand, there are several major *thermal* processes that have been used for magnesium production over the past period. A reducing agent, such as silicon, is used to reduce magnesium oxides under a vacuum. The magnesium oxides are derived basically from dolomite ($MgCO_3 \cdot CaCO_3$) and magnesite ($MgCO_3$).

1.4 The Establishment of the International Society for the Development of Research on Magnesium

The period known as the development in the field of chemical technology extended all the way up to the first quarter of our 20th century. The First International Symposium on magnesium as well as the creation of SDRM* (the International Society for the Development of Research on Magnesium), have contributed to the present modern period, which began in 1971.

SDRM is recognized as an international coordinating body.

There is a great deal of research and pilot plant work being done at present in response to the demand on magnesium products. Technical skills and advocated type of researches are being applied to magnesium now more than ever before at any time in the history. It was reported that in the United States, 63% of magnesium production came from seawater and brines during 2015. While this may sound like a good number, the production of magnesium is neither cheap nor clean. When a survey was carried out on the number of methods used to produce magnesium, it was found that these methods vary over a wide range. They include methods ranging from electrolytic processes all the way to high temperature reduction, which is known as the *Pidgeon* method. The latter types are identified as energy-intensive processes. It has been reported that 10 kg of coal is consumed as energy input in the production of one Kg of magnesium by using the Pidgeon process.

This is the full historical text as reported about the establishment of SDRM[*]:
The International Society for the Development of Research on Magnesium (SDRM)
is a non-profit organization. The purpose of the Society is to promote magnesium
research in all branches of life science and medicine by increasing the collabora-
tion, diffusion and exchange of information. SDRM holds international scientific
meetings and supports the publication and presentation of scientific results. It was
founded in the early '70s in Paris by a group of medical doctors under the leader-
ship of Prof. Jean Durlach, who foresaw how magnesium status could be crucial in
a wide variety of relevant pathologies ranging from cardiovascular to metabolic and
neurological diseases.

1.5 Minerals Cycle

In the early stages of the industrial development of the world, the normal
cycle of handling a resource mineral was to take the most concentrated
form of deposits available, purify it, and use it. Next, allow for that mineral to return to nature once more. The ways it can be returned to the
universe are so diverse to the point that it is doubtful that such mineral be
recovered again.

As an example, let us consider the case of *iron*. It is produced from the rich-
est ore deposits and made into different useful objects by mankind, which
immediately begin to rust away upon usage. Obviously, using iron in this
fashion will lead to deconcentration of the ore to the point where it is scat-
tered beyond use.

Looking at magnesium, it has been found that after only a few years of
commercial production, this type of metal behaves in a completely different way. It is being produced from the most diluted form in which it
is found in nature. That is a source which is also the most widespread
throughout the world.

The development of the magnesium industry is the most outstanding
illustration of how an industry should grow. Here we have a metal produced from a source widespread throughout the world, a source that is
inexhaustible, right at the start of the development of its cycle.

If we were to plan our ideal metal, its properties is not all what we are
looking for. This ideal behavior of magnesium does not exist in any other
metal. But this would make the guidelines on how to handle magnesium,
alloying it with other metals in order to produce the physical and chemical
properties we want.

---

[*] https://www.acronymfinder.com/SDRM.html.

1.6 The Motivation for Writing a Book on Magnesium

To the author's surprise, very few textbooks could be found on the extraction and the recovery of magnesium from seawater. This is one good reason that motivated the writing the of text "Magnesium from Resources to Production." The other reason is the author's involvement in experimental research-work in this field for a period of more than 30 years. The author and coworkers initiated work to test the feasibility of applying preferential salt separation (PSS) to recover $MgCl_2$ from seawater. This is a physical method based on the successive precipitation of soluble salts in evaporating seawater flowing along an inclined evaporator (fully explained in Chapter 5).

This published work in this area attracted the attention of some industrial organizations to consider using the PSS to build pilot units for the production of magnesium south of Italy (correspondences with the project manager in charge, at that time, are found in Appendix C).

It should be stated that using solar energy in the experiments is a cornerstone in this work. Advantages are illustrated in the conceptual diagram shown in Figure 1.4.

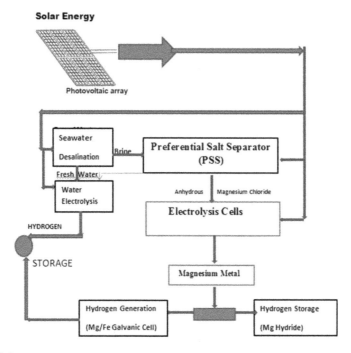

FIGURE 1.4
Coupling of solar energy with PSS.

In conclusion, one can postulate that the economics of magnesium could be very well enhanced by considering two important factors:

- Development in the extraction technology of magnesium
- Finding new avenues for technical application of magnesium to provide power for the energy domain

Finally, magnesium is best described as "the light metal with solid industrial markets and great future potential" which is 75% lighter than steel and 33% lighter than aluminum.

2

Occurrence and Resources of Magnesium

2.1 Background

An *ore* is defined as a type of rock that contains sufficient minerals with important elements. Usually, ores include metals that can be economically extracted from the rock. By using mining techniques, the ores are extracted from earth. Next, they are refined to extract the valuable elements contained in the ore. Usually smelting is used as the refining process.

The costs attained for mining the ore are normally a function of:

- The grade or concentration of an ore mineral, or metal
- Its form of occurrence

The metal value contained in the rock has to be balanced against its cost of extraction. This is important in order to decide:

- What type of ore can be processed?
- What ore is of too low a grade to be worth mining?

Metal ores normally include oxides, sulfides, and silicates. Others are known as *native* metals (such as native copper). They are not commonly concentrated in the earth's crust. On the other hand, *noble* metals (not usually forming compounds) such as gold are found in the earth's crust. Processing the ores is a must in order to extract the metals contained in these ores from the waste rock. An ore deposit designates an occurrence of a particular ore type. Most ore deposits are named after their location.

2.2 Occurrence

Magnesium occurs and is found in many places and many locations:

- It is the *eighth* most abundant element in the earth's crust and the *fourth* most common element in the earth, following iron, oxygen, and silicon.
- It is the *third* most abundant element dissolved in seawater, where sodium and chlorine are ranked first and second, respectively.
- In addition, about 13% of the planet's mass is made by magnesium, and about 2.1% of the earth's crust is composed of magnesium, which is a sign of its abundance.

Magnesium is also found in seawater, occurs naturally, and is found strictly in combination with other elements. It has a +2 oxidation state.

- Magnesium is also abundant in the terrestrial crust. Among the ore minerals, the most common are the carbonates magnesite (magnesium carbonate $MgCO_3$) and dolomite (a compound of magnesium and calcium carbonates, $MgCO_3 \cdot CaCO_3$). It occurs as kieserite, ($MgSO_4 \cdot H_2O$), schonite ($K_2SO_4 \cdot MgSO_4 \cdot 6H_2O$), kainite ($MgSO_4 \cdot KCl \cdot 3H_2O$), and carnallite ($MgCl_{12} \cdot KCl \cdot 6H_2O$) in the salt beds at Strassfurt in Germany.
- Magnesium also occurs as soluble salts including *Epsom* salts, which are found in mineral waters and hard waters.
- Magnesium is found as its oxide, being a constituent of rocks in different ore forms that include:
  - Igneous rocks, such as olivine, $(MgFe)_2SiO_4$.
  - Sedimentary rocks, such as the carbonate, magnesite, and the mixed carbonate, dolomite.

Turkey, North Korea, China, Slovakia, Austria, and Russia are the largest producers, worldwide, of magnesium ores.

2.3 Sources of Magnesium

Sources of magnesium are identified as shown in Figure 2.1: These magnesium ores are pictorially illustrated in Figure 2.2 (after Asian metal).

Globally, resources for Mg could be described as *virtually* unlimited. It is abundant and found in dolomite and magnesium-bearing evaporite minerals, which are plentiful. Brines, on the other hand, are also rich in

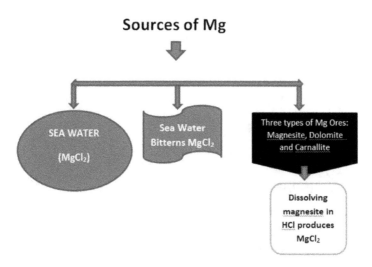

FIGURE 2.1
Sources of magnesium.

FIGURE 2.2
Illustration of the different types of magnesium ores. (From Asian metal, The World Metal Information Center, www.asianmetal.com, accessed October 6, 2015.)

magnesium. They constitute a huge resource, estimated to be in the billions of tons. Finally, and most importantly, is the recovery of Mg from seawater from many countries located at coastlines.

The following table designates three separate sources of magnesium from seawater:

Location	Concentration of Mg	Type of Source
Seashore	1,350 mg/L	Seawater
Typical oceans	56,000 mg/L	Seawater bittern (SWB)
Dead Sea	35,200 mg/L	Saline lake

Russia, China, and Korea are the countries that have most of the magnesium reserves in them. Magnesite represents the major source of magnesium for these countries. The identified world resources of magnesite total 12 billion tons. On the other hand, world reserves of magnesite total 2.4 billion tons. China's share in this reserve accounts for 21%, which is about 500 million

FIGURE 2.3
Magnesium sources in Dead Sea and Great Salt Lake. (From Asian metal, The World Metal Information Center, www.asianmetal.com.)

tons; Russia comes next 27%, about 650 million tons; and North Korea comes in third with 18.8%, about 450 million tons.

In addition to magnesite resources, well and lake brines, and also seawater, represent vast reserves from which magnesium compounds can be recovered. The Dead Sea and the Great Salt Lake (located in Utah in the United States) represent abundant resources of magnesium as illustrated in the map shown in Figure 2.3.

Natural mineral deposits of magnesia-potassium salts are the principal magnesium ores. In addition, metamorphosed dolomites represent large magnesite beds. In cases of contact metamorphism of magnesite, bodies of brucite (the raw material with the highest magnesia content) are formed.

When leaching of magnesia salts by subterranean waters occurs, natural mineral brines and salt springs are formed. On the other hand, modern salt deposits (brines and sediments) can also occur in inland gulfs and in intracontinental drainage basins, for example, the Great Salt Lake in the United States. Another increasingly used magnesium source is seawater (4% Mg in dry residue) with its stable composition and unlimited resources.

The principal magnesium ores are found in mineral deposits of magnesia-potassium salts. Large magnesite beds are found in metamorphosed dolomites. Bodies of brucite (the raw material with the highest magnesia content) form in cases of contact metamorphism of magnesite. Natural mineral brines and salt springs are formed as a result of leaching of magnesia salts by subterranean waters. Modern salt deposits (brines and sediments) occur in inland gulfs and in intra continental drainage basins, for example the Great Salt Lake in the USA.

Large magnesia-potassium salt basins are found in some parts of Russia. Other well-known areas are the Permian Stassfurt salt basin (Federal Republic of Germany and German Democratic Republic) and deposits in the southern United States.

2.3.1 Magnesite

Magnesite, in the form $MgCO_3$, represents an important ore used for magnesium production. It is the source of a range of industrial minerals. If pure magnesite is found, it contains 47.8% magnesium oxide and 52.2% carbon dioxide. However, natural magnesite almost always contains other compounds. This may include calcium carbonate as the mineral calcite, and iron carbonate as the mineral siderite.

The world magnesite reserves are shown in Table 2.1 with a total of 2.4×10^6 tons.

The list of countries who were magnesite producers in 2014 is given in Table 2.2.

Another compound for magnesium is dolomite, which has the formula $CaMg(CO_3)_2$. In the dolomite, $MgCO_3$ constitutes 45.65% and $CaCO_3$ constitutes 54.35%. The color of magnesite is one of its distinguishing properties. It occurs in a wide range of sands, and the color depends on the concentration. The color, therefore, varies from white when pure to yellowish or grey, white, and brown. Hardness is another important property. On the Mohs, it is in the range of 3.5–4.5. Similarly, the specific gravity varies from 3–3.2.

Large magnesite beds are found in metamorphosed dolomites. Bodies of *brucite* (the raw material with the highest magnesia content) are formed in some locations. Natural mineral brines and salt springs are formed as a result of leaching of magnesia salts by subterranean waters. Modern salt deposits (brines and sediments) occur in inland gulfs and in intracontinental drainage basins. A good example is the Great Salt Lake in the USA.

TABLE 2.1

World Magnesite Reserves (100 metric tons of Magnesium Content)

Country	Reserves
United States	10,000
Australia	95,000
Austria	15,000
Brazil	86,000
China	500,000
Greece	80,000
India	20,000
Korea, North	450,000
Russia	650,000
Slovakia	35,000
Spain	10,000
Turkey	49,000
Other countries	390,000
World total (rounded)	2,400,000

Source: US. Geological Survey (GS), 2014. https://www.usgs.gov/, https://www.usgs.gov/locations/us-geological-survey-headquarters.

TABLE 2.2

List of Countries by Magnesite Production 2014

Rank ⬥	Country/Region ⬥	Magnesium production production (thousand tonnes) ⬥
—	*World*	6,970
1	China	4,900
2	Russia	400
3	Turkey	300
4	Austria	200
5	Slovakia	200
6	Korea, North	80
7	Brazil	150
8	Spain	280
9	Greece	115
10	India	60
11	Australia	130
—	Other countries	155

Source: U.S. Geological Survey. Retrieved September 15, 2016.

Some areas in Russia have very large magnesia-potassium salt basins. Other well-known areas are the Permian Stassfurt salt basin (Federal Republic of Germany and in the southern USA).

In seawater and lake brines, magnesite and dolomite are sources of magnesium. Since water is abundant, it is considered its number one source.

Two physical forms of magnesite exist: cryptocrystalline or amorphous magnesite, and macrocrystalline magnesite. Magnesite can be formed in five different ways:

1. As a replacement mineral in carbonate rocks
2. As an alteration product in ultramafic rocks (igneous rocks composed mainly of one or more dark-colored ferromagnesian minerals)
3. As a vein-filling material, or a sedimentary rock
4. As nodules formed in a lacustrine (lake) environment
5. As a replacement-type magnesite deposit involving magnesium-rich fluids entering limestone via openings to produce both magnesite and dolomite

The action of carbon dioxide-rich waters on magnesium-rich serpentinite rock creates alteration-type deposits. These rocks have been formed from the alteration of magnesium and iron silicate minerals. The resulting magnesite may be very pure.

The formation of sedimentary deposits is different. They usually occur as thin layers of variable magnesite quality.

2.3.2 Mining and Extraction

The way mining and extraction are done for the ores is different for each type:

- Conventional methods for mining are applied for dolomite and magnesite, followed by concentration.
- Carnallite, on the other hand, is treated differently. Solution mining is required for this type of magnesium. It is dug up as ore or separated from other salt compounds that are brought to the surface using the solution mining technique.
- Naturally occurring magnesium-containing brines are concentrated in large ponds by solar evaporation.

In general, the following are guidelines for a feasible successful extraction of an ore deposit to be carried out:

- Prospecting or exploration to first find, then define, the extent of the discovery. Valuation is carried out for the ore where it is located. This is simply referred to as the *ore body*.
- Size estimation is done for the ore resource by mathematical methods, followed by determining the grade-type of the ore at hand.
- A theoretical study of the economics of the ore deposit is conducted based on a *pre-feasibility* study. This step helps the investor make a decision as early as possible: should further investment in estimation and engineering studies be carried out? The answer could be either yes or no. If no, the project is terminated. If yes, this would back up and stand for risks that may be involved.
- A further complete feasibility study is carried out to bug the project completely.
- Part of the early plant design is to plan to create access to the plant location for an ore body on-site in order to accommodate all equipment needed for the plant as well as other facilities.

The feasibility study is normally a part of plant design. A course on this subject is devoted by chemical engineers. This is referenced briefly in Figure 2.4. In this figure, all the steps involved in a design project are illustrated.

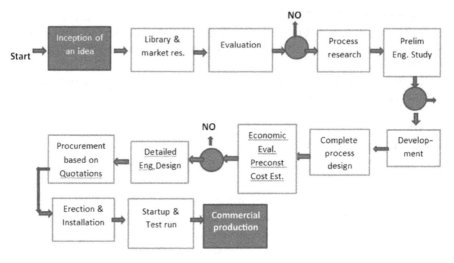

FIGURE 2.4
The A–Z chart in plant design.

Since magnesium forms stable compounds and reacts with oxygen and chlorine in both the liquid and gaseous state, a strong chemical reagent is needed for the extraction process of magnesium. This means that extraction of the metal from raw materials is an energy-intensive process. It requires well-tuned technologies.

Once magnesium compounds are made available, commercial production of the elementary metal can follow two completely different methods: electrolysis of magnesium chloride, or thermal reduction of magnesium oxide. If the cost of electric power is low, electrolysis is the method to be applied. This approach of magnesium chloride electrolysis represents approximately 75% of world magnesium production, which will be fully explained in next chapters.

2.4 Magnesium Resources World Wide

2.4.1 Australian Resources, Deposits, and Processing

Australia has a proven reserve of magnesite, estimated to be 202 million tons. *Proven reserve* means that such a reserve could be economically extracted at current prices with existing technology.

On the other hand, the estimated world economic resources (proven reserve) of magnesite are about 8,600 million tons of $MgCO_3$. In this regard, China comes first in terms of its huge reserve, followed next by the Russia and North Korea. Historically, these deposits have been formed by the deposition in lakes of magnesium bicarbonate, which is derived from the alteration of serpentinite rock. Next comes evaporation, which is behind the precipitation of hydrated magnesium carbonate. The formation of hard nodules of dehydrated magnesite was done by the effect of the deposition of mud over the magnesite, followed by further evaporation. The mining of these deposits dates back to the year 1989.

In New South Wales, the magnesite ore contains 95%–99% $MgCO_3$. Its thickness varies between 2–10 m. Other occurrences of magnesite in New South Wales are known to be at Lake Cargelligo and Cobar.

Beds of magnesite ranging from 5 cm to 9 m in thickness are found in South Australia. These deposits occur in Witchelina, 80 km northwest of Leigh Creek, Copley, and Myrtle Springs. Similarly, nodules in dark clayey material crop are reported to exist in Western Australia.

The processing of magnesite involves different courses. The following are typical examples encountered in more than location:

- When crude magnesite is heated to between 700°C–1000°C, carbon dioxide is driven off to produce caustic-calcined magnesia (caustic magnesia). This type of magnesite is characterized as being able to absorb liquids, heavy metals, and ions, as well from liquid streams. This type of magnesite turns out to be used making in applications of water treatment.

- When calcined magnesia is heated to a higher temperature in the range of 1530°C–2300°C, it produces nonreactive products. As a result of this heating, exceptional stability and strength at high temperatures is exhibited. The product produced is known as *dead-burned*. It is also known as *sintered* magnesia and is mainly used as a refractory material.

- Lastly, when calcined or dead-burned magnesia is subjected to a temperature in excess of 2800°C using an electric arc furnace, electrofused magnesia is obtained. This compound has higher strength, resistance to abrasion, and chemical stability, which makes it superior dead-burned magnesia. As an application, premium-grade refractory bricks are manufactured using this compound. These bricks find many other applications; for example, it is used in high-wear hot-spots of basic oxygen furnaces, such as the electric arc. It is most needed for furnaces where temperatures can approach 950°C.

Another source for magnesia is the processing of seawater and magnesium-rich brines. This is not a promising way to obtain magnesium. It is a rather a complex process that consumes much more energy than simply heating natural magnesite.

As far as magnesium production, currently Australia does not produce magnesium metal. However, in 1998, the Australian Magnesium Corporation commenced operating a 1,500 ton per year demonstration plant at Gladstone in Queensland. It plans to establish a commercial magnesium metal plant at Gladstone using the Australian Magnesium Process.

2.4.2 Magnesium Resources in China

China is the country with the second-most abundant magnesite resources. A minimum proven reserve of 350 million tons exists in many regions of the country. The reserves in Liaoning are the largest, accounting for about 85.6% of those in China. It is described as a world-famous area for its large reserves along with high-quality magnesite.

China also rich in dolomite ores. The proven reserve of dolomite in China is estimated to be nearly 4 billion tons. Accordingly, dolomite mines are plentiful. They are mainly being distributed in many provinces like Shanxi and Henan.

In addition, magnesium salt is abundant in Qaidam Basin in China. It represents about 99% of identified magnesium salt reserves that are found in the country. There are two types of reserves:

- Magnesium chloride, with proven (identified) reserves of 4.2 billion tons and a base reserve of 1.9 billion tons
- Magnesium sulfate, with proven (identified) reserves of 1.7 billion tons, and a base reserve of 1.2 billion tons.

The world's magnesite mine production was reported to drop to 5.96 million tons in 2013, based on the latest data released by USGS in 2013. In addition, the report includes a year-on-year decrease of 390 tons. At the same time, it was stated that China tops the world's magnesite production. China accounts for 57% of the total figures released.

In 2010, China produced over 3.2 million metric tons of the carbonate.

Two-thirds of worldwide magnesite reserves are concentrated in China. Turkey ranks second with an annual production of 6 million tons of magnesite. The third place is occupied by North Korea, followed by Russia, Austria, Slovakia, Spain, Greece, India, Brazil, and Australia.

Magnesite finds a variety of industrial and agricultural applications:

- It is used as a feed supplement to cattle, and is found in fertilizers.
- It is used as an insulating material in electrical appliances, industrial fillers.
- It is used to desulfurize flue gas, and to get rid of sulfur compounds.

2.5 Primary Magnesium Production

Basically, most of the world's primary magnesium is produced in China. The production in 2013 was estimated to be 800,000 tons, while the total world production was 910,000 tons. Other significant producers of magnesium

TABLE 2.3

World Primary Magnesium Production (In 103 metric tons of Magnesium Content)

Country	2012	2013
United States	W	W
Brazil	16	16
China	698	800
Israel	27	28
Kazakhstan	21	21
Korea, Republic of	3	9
Malaysia	5	5
Russia	29	30
Serbia	2	2
Ukraine	2	2
World total (rounded)	802	910

Source: US. Geological Survey (GS), 2014.

metal are Russia, Israel, and Kazakhstan, with smaller quantities are produced in Canada, Brazil, Serbia, and the Ukraine, as given in Table 2.3.

2.6 "Extraction" is a Key Process

There is a good reason to call extraction a key process that makes the elementary magnesium metal available for mankind from these resources.

It is helpful to recognize the meaning of the word "to extract." It simply means to remove or take out, especially by effort or force. In this context, extraction of magnesium can be done through two different routes:

First—Extraction of magnesium from its ores using a suitable process:

Here we have more than one case. To extract magnesium from dolomite, for example (an ore containing magnesium and calcium), one needs to make sure that calcium is not precipitated at the electrodes by maintaining the optimum voltage through the use of energy.

For extraction in general, magnesium is known to form stable compounds and reacts with oxygen and chlorine in both the liquid and gaseous state. This means that extraction of the metal from raw materials is an energy-intensive process requiring well-tuned technologies. Commercial production follows two completely different methods: electrolysis of magnesium chloride or thermal reduction of magnesium oxide. Where power costs are low, electrolysis is the cheaper method. It accounts for approximately 75% of world magnesium production.

Second—Extraction of magnesium from seawater:

It is well known that the Mg^{2+} cation is the second-most abundant cation in seawater, because its mass is about one-eighth of the mass of sodium ions in a given sample. This makes seawater and sea salt attractive commercial sources for the recovery of magnesium.

To extract the magnesium from seawater and brines, calcium hydroxide is added to seawater to form magnesium hydroxide precipitate. It is then separated and converted to $MgCl_2$ by adding HCl. We consider this a preliminary stage to capture the magnesium ion and convert it to $MgCl_2$. We notice that extraction took place by using the chemical energy of the formation spent in the reactions. We call this process the chemical approach of obtaining $MgCl_2$.

The proposed method (PSS), on the other hand, is based on the physical separation of $MgCl_2$. By using this approach, energy in the form of solar heat is utilized to evaporate seawater to obtain $MgCl_2$ as an end product.

Next, to obtain magnesium as end product, that is to extract magnesium from $MgCl_2$, again we have to spend energy to liberate the magnesium from $MgCl_2$. This is done by the well-known electrolytic process. The sequence of the extraction steps leading to the full recovery of magnesium metal is illustrated in Figure 2.5.

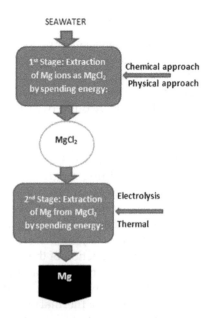

FIGURE 2.5
Extraction of magnesium metal from seawater.

To summarize, the extraction process of magnesium as end product, that requires spending energy, consists of the following two steps:

1. Extraction of magnesium ions from seawater by either chemical reactions (Dow Process) or by the physical separation (PSS proposed process), to form $MgCl_2$.

2. Extraction of magnesium metal from $MgCl_2$ by the electrolytic process or the thermal process.

3

Magnesium and Mineral Salts from Seawater: Exploration Study

3.1 Introduction

The concentration of magnesium in seawater is about 0.13%. It is the most commonly found cation in oceans, after sodium. On the other hand, magnesium is found in a large number of minerals. Typical examples are dolomite (calcium magnesium carbonate; $CaMg(CO_3)_2$) and magnesite (magnesium carbonate; $MgCO_3$).

The source of magnesium is rocks. When washed away by rain from rocks, it ends up in seawater.

Seawater is a source of minerals that have extracted and mined since ancient times. At the top of the list comes sodium chloride, which extracted in ancient times in China in particular. At present, the following four metals are the ones extracted:-

- Sodium, Na
- Magnesium, Mg
- Calcium, Ca
- Potassium, K

These metals are commercially extracted in the form of Cl^-, SO_4^{2-}, and CO_3^{2-}. In addition, Mg is also extracted as MgO. On the other hand, it has been found that the recovery of mineral elements from seawater with low concentrations did not receive much attention. This is because of an obvious economic reason: the market values of such metals are much lower than the total production costs, which is the sum of capital and operational costs spent in the extraction process.

It is estimated that over 40 minerals and metals are found in seawater. The potential extraction of such minerals may receive much attention in the future. Other source and references report 47 minerals and metals in seawater (reported by Mining Weekly, April 1, 2016, Creamer Media). Among these metals is the strategic energy metal, *uranium*. It is estimated that the world's oceans contain 4.5 billion tons of uranium metal.

Knowing these minerals and metals are found in seawater is one thing; extracting them is quite another. Yet minerals and metals are being extracted from seawater, and inland briny waters, on a commercial basis. One of them is so obvious in our daily use that it is often forgotten: salt, or sodium chloride. While consuming too much salt can be bad for your health, consuming some salt is essential to stay healthy and alive. Salt is also used to treat ice-covered roads in many, mostly northern hemisphere, countries in winter, and it serves as a raw material for the manufacture of hydrogen, chlorine, and sodium hydroxide through the process of electrolysis. In fact, 68% of all salt produced is used in manufacturing and industrial processes. In all, salt has about 14,000 different applications.

3.2 Principal Mineral Resources

3.2.1 Historical Background

It is well established that oceans host a vast variety of geological processes responsible for the formation and concentration of mineral resources. They cover a vast area of earth's surface, about 70%. Many materials eroded or dissolved from the land surface come from oceans as the ultimate source. Consequently, oceans are regarded as *storehouses* that contain vast quantities of materials; at present, we enjoy using them as major resources for humans. Currently, the extraction of resources from oceans is kind of a limited process. Only the following are currently extracted from oceans: salt; magnesium; placer gold, tin, titanium, diamonds, and fresh water.

When we look for ancient ocean deposits of sediments and evaporites now located on land, it is found that they were originally deposited under variable marine conditions. Such deposits represent an attractive route for extraction, and have the potential to be exploited on a very large scale. This is because of easier accessibility and lower cost of terrestrial, compared to modern marine resources.

Natural gas and oil resources have been extracted from the seas for a long time (Hussein K. Abdel-Aal, et al., 2015). On the other hand, ores and mineral deposits on the sea floor were left unutilized. They could not attract the interest of the explorer. With the increase in resource prices, the incentive in ocean mining increased. The mining of some specific ores like massive sulfides and manganese nodules is expected to gain attention within the next few years. Commercial interest in manganese nodules and cobalt crusts is rising.

As a result of the rising interest in marine mining, the International Seabed Authority (ISA) has been established in Jamaica. In addition, the United Nations Convention on the Law of the Sea (UNCLOS) was signed in 1982. This marks what is called the *constitution for the seas* which came into effect in 1994. The basis for signatories' legal rights to use the marine resources on the sea floor—outside national territorial waters—were constituted, based on the rules of this major convention. Once the prices dropped internationally, the industrial countries lost interest in resources. From the economic point of view, it became no longer profitable to retrieve the accretions from the deep sea. Utilization of the metals they contained came to halt. This went on concurrently with the discovery of new onshore deposits that were cheaper to exploit.

Two factors are behind the present resurgence of interest:

First: The sharp increase in resource prices.

Second: The attendant rise in profitability of the exploration business. This is triggered by strong economic growth in countries like China and India. They purchase large quantities of metal on world markets. Geopolitical interests in safeguarding these supplies of resources also play a role. In addition to the fact that there has been an increasing demand for resources, countries with no reserves started seeking to assert extraterritorial claims in the oceans.

3.2.2 Minerals Extracted

The principal mineral resources presently being extracted and likely to be extracted in the near future are:

- Salt
- Potassium
- Magnesium
- Sand and gravel
- Limestone and gypsum
- Manganese nodules
- Phosphorites
- Metal deposits associated with volcanism and seafloor vents

A few of these will be examined and described next.

a. *Salt*: Sodium chloride, or salt, is found in seawater at a concentration in the range of 3–5%. More than 80% of the total dissolved chemical elements in seawater are made up by sodium chloride. It has been estimated that the total mass of salt available in all the oceans is huge and more than enough. It could supply all human needs for hundreds of years. Most of the nearly 200 million metric tons of salt produced annually is mined from large beds of salt. In addition, it is the practice of many salt-production facilities to harness salt and extract it directly from the oceans simply by direct evaporation of water subjected to open solar radiation. The residual salts will separate and be collected.

b. *Potassium*: Although potassium occurs in vast quantities in seawater, we find that its average concentration is rather low; about 1,300 parts per million (0.13%). This is such a low concentration that direct economic extraction is uncertain. On the other hand, potassium salts are found along with common salt in salt beds. Mining from these beds takes place at a commercial scale in the range of tens of millions of metric tons per year.

c. *Magnesium*: It is the only metal directly extracted from seawater. In seawater, magnesium concentration is about 1,300 parts per million (0.13%). At present, an appreciable quantity of the magnesium metal and many of the magnesium salts produced in the United States are extracted from seawater (about 60%–70%). Magnesium ores (rocks) from ocean deposits are the source for the raw materials that produce additional magnesium metal. The principal minerals mined in this regard are magnesite and dolomite.

d. *Water*: Oceans represent a total volume of more than 500 million cubic kilometers, representing more than 97% of all the water on Earth. Seawater is salty, with about 3% salt content. This makes it unusable for drinking and for most applications of human needs.

Desalination of seawater to obtain fresh water from ocean water was the only outlet practiced by many countries for a long time. This source is limited in volume, besides being expensive compared to other sources of fresh water. The search for other methods to obtain fresh water from seawater is in progress. Reverse osmosis offered a new approach to increase obtaining fresh water from seawater at a much high efficiency.

Consideration should be given to such factors as geographic locations and availability and cost of energy resources when large-scale production is sought.

3.3 Minerals in Sea Salt

Unlike table salt, sea salt is created by evaporating seawater and contains more minerals than other salts because it comes from the sea. It involves little or no processing and contains no anti-clumping additives. The color and the flavor that belongs to sea salt are attributed to the minerals found in it.

Sea salt stands on a different plateau. It poses some advantages because of their different properties. It contains major and minor elements that can support the human body's immune system and aid in the body's normal growth and development. This is biologically explained by the fact that the body of a human being needs minerals to function properly. Such minerals are needed to support the immune system. In addition, they aid in body's normal growth and development. Clinically, major minerals are minerals the body needs in quantities of about 100 milligrams per day.

Sodium, potassium, phosphorus, and calcium are recognized as major minerals found in sea salt. Their functions are described as follows:

Sodium helps balance the body's fluids and is needed for proper muscle contraction. Potassium, on the other hand, helps maintain a steady heartbeat and aids in transmitting nerve impulses. Phosphorus and calcium are crucial in the development and protection of strong bones and teeth. The body also requires trace minerals in very small amounts. Depending on the source, important trace minerals found in sea salt include iron, iodine, manganese, and zinc. Iron helps red blood cells and muscle cells carry oxygen throughout your body. Iodine is related to the thyroid hormone and aids in regulating your body's temperature. Manganese contributes to proper bone development, and it also aids in the metabolism of amino acids and carbohydrates. Zinc plays an important role in developing new cells and in healing wounds.

Open ocean water contains dissolved salts in a range of 33–37 grams per liter, corresponding to a total mass of about 10^{16} tons. It is a huge amount to be invested.

The oceans contain these immense amounts of dissolved ions which, in principle, could be extracted without the complex and energy-intensive processes of extraction and beneficiation that are typical of land mining. Concentrations and estimated amounts of dissolved metal ions in the sea are compared with the estimated land resources in Table 3.1. The amount of magnesium found in ocean is 10^6 times that of the land reserve.

The four most concentrated metal ions, Na^+, Mg^{2+}, Ca^{2+}, and K^+, are the only ones commercially extractable today, with the least concentrated of the four being potassium (K) at 400 parts per million (ppm).

Mineral make up of seawater is reported as follows (in order from most to least):

Element	Molecular Weight	PPM in Seawater	Molar Concentration
Chloride	35.4	18980	0.536158
Sodium	23	10561	0.459174
Magnesium	24.3	1272	0.052346
Sulfur	32	884	0.027625
Calcium	40	400	0.01
Potassium	39.1	380	0.009719

(Continued)

Element	Molecular Weight	PPM in Seawater	Molar Concentration
Bromine	79.9	65	0.000814
Carbon (inorganic)	12	28	0.002333
Strontium	87.6	13	0.000148
Boron	10.8	4.6	0.000426
Silicon	28.1	4	0.000142
Carbon (organic)	12	3	0.00025
Aluminum	27	1.9	0.00007
Fluorine	19	1.4	0.000074
N as nitrate	14	0.7	0.00005
Nitrogen (organic)	14	0.2	0.000014
Rubidium	85	0.2	0.0000024
Lithium	6.9	0.1	0.000015
P as Phosphate	31	0.1	0.0000032
Copper	63.5	0.09	0.0000014
Barium	137	0.05	0.00000037
Iodine	126.9	0.05	0.00000039
N as nitrite	14	0.05	0.0000036
N as ammonia	14	0.05	0.0000036
Arsenic	74.9	0.024	0.00000032
Iron	55.8	0.02	0.00000036
P as organic	31	0.016	0.00000052
Zinc	65.4	0.014	0.00000021
Manganese	54.9	0.01	0.00000018
Lead	207.2	0.005	0.000000024
Selenium	79	0.004	0.000000051
Tin	118.7	0.003	0.000000025
Cesium	132.9	0.002	0.000000015
Molybdenum	95.9	0.002	0.000000021
Uranium	238	0.0016	0.0000000067
Gallium	69.7	0.0005	0.0000000072
Nickel	58.7	0.0005	0.0000000085
Thorium	232	0.0005	0.0000000022
Cerium	140	0.0004	0.0000000029
Vanadium	50.9	0.0003	0.0000000059
Lanthanum	139.9	0.0003	0.0000000022
Yttrium	88.9	0.0003	0.0000000034
Mercury	200.6	0.0003	0.0000000015
Silver	107.9	0.0003	0.0000000028
Bismuth	209	0.0002	0.00000000096
Cobalt	58.9	0.0001	0.0000000017
Gold	197	0.00008	0.00000000004

Source: https://web.stanford.edu/group/Urchin/mineral.html.

TABLE 3.1

Mineral Reserves in Ocean and Land

Element	Concentration in Seawater (ppm)	Total Oceanic Abundance (tons)	Mineral Reserve on Land (tons)
Na	10,800	1.40×10^{16}	–
Mg	1,290	1.68×10^{15}	2.20×10^{9}
Ca	411	5.34×10^{14}	–
K	392	5.10×10^{14}	8.30×10^{9}

Source: Bardi, U., *Sustainability*, 2, 980–992, 2010.

3.4 Evaporation and Separation Sequence of Salts

Magnesium (Mg) exists in seawater as ions of magnesium. When it comes to the mineral salts found in seawater, the major components of natural seawater determining their solubility properties are: Na^+, K^+, $Mg_2{}^+$, Cl^-, and SO_4, as shown in Figure 3.1.

According to Balarew (1993), it is usually assumed that the major constituents present in the sea show constant relative proportions. He further added that the variations in the composition of waters from different seas are due only to the changes in the amount of water present. Accordingly, one can conclude that the composition-density diagram shown in Figure 3.2 for the initial composition of the Black Sea water is valid for every seawater type.

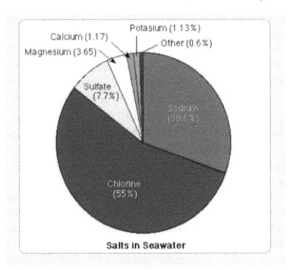

FIGURE 3.1

Major components in natural seawater.

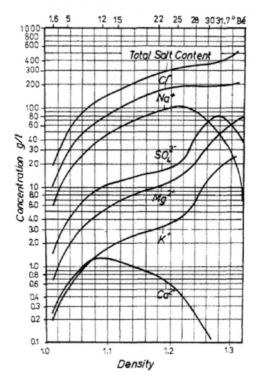

FIGURE 3.2
Composition-density changes during evaporation of seawater.

When seawater (composition is given in Table 3.2) is evaporated, soluble salts will be formed at different stages during the evaporation. The crystallization of the salts dissolved in seawater is governed by their solubility products and occurs at different concentration levels. When seawater is concentrated gradually, brine concentration increases, leading to the successive precipitation of the least soluble salts first. For example, iron oxide and calcium carbonate start to crystallize first, in very small quantities, followed by calcium sulfate (known as gypsum).

The work published by manoa.hawaii.edu and shown in Figure 3.3 identifies the salt rings formed when seawater was evaporated from a watch glass. The outer ring, which precipitated out of solution first, is primarily made up of calcium carbonate ($CaCO_3$). Carbonates are the least soluble salts in seawater. The inner ring is primarily made up of potassium (KCl) and magnesium ($MgCl_2$) salts, which are very soluble.

This type of experiment, which was carried out using a watch glass, is very significant and informative when it comes to the presentation on the

TABLE 3.2

Composition of Seawater

Components	Amount Present (g/1000 g) Seawater
Ca	0.408
SO_4	2.643
Mg	1.265
Cl	18.95
K	0.38
Na	10.48
Br	0.065
Total	**34.19**

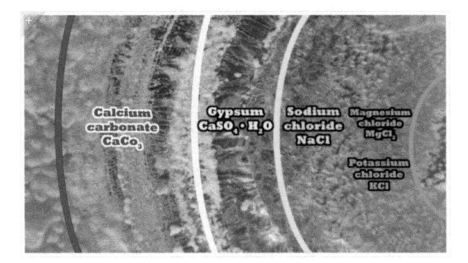

FIGURE 3.3
Identification of salts by evaporating seawater on watch glass: black ring is the least soluble and gray ring is the most soluble. (From Exploring Our Fluid Earth. Weird Science: Types of Salts in Seawater, https://manoa.hawaii.edu/exploringourfluidearth/chemical/chemistry-and-seawater/salty-sea/weird-science-types-salts-seawater.)

Preferential Salt Separation, PSS, for the extraction of magnesium from seawater in Chapter 5. This experiment is supportive of the basic principle behind the PSS; that is, magnesium chloride is the last salt to precipitate from evaporating seawater.

The solubility of each of the main salts that precipitate out of seawater is shown in Table 3.3. The taste of these salts varies based on their chemical composition.

TABLE 3.3

Solubility of Salt Compounds in Seawater

Salt Compound	Picture	Solubility	Taste
$CaCO_3$ (calcium carbonate)	Image courtesy of Ondrej Mangl	Insoluble at 50% evaporation ($\approx$70 ppt)	Chalky
$CaSO_4 \cdot H_2O$ (gypsum)	Image courtesy of Wilson44691	Insoluble at 80% evaporation ($\approx$100 ppt)	Chalky
NaCl (sodium chloride or halite)	Image courtesy of NASA	Insoluble at 90% evaporation ($\approx$130 ppt)	Salty
KCl and $MgCl_2$ (potassium and magnesium salts)	Image courtesy of Romain Behar	Insoluble at >95% evaporation (>150 ppt)	Salty

Source: Exploring Our Fluid Earth. Weird Science: Types of Salts in Seawater. https://manoa. hawaii.edu/exploringourfluidearth/chemical/chemistry-and-seawater/salty-sea/ weird-science-types-salts-seawater.

3.5 Theory of Evaporating Seawater and Precipitating Salts

3.5.1 Ocean Water Profile

It is necessary to describe the ocean's water profile before we get to discuss our main theme of evaporating seawater. It is true that this applies for layers of Deep Ocean Water (DOW), but the phenomena are worth knowing.

There are three distinctly different layers of ocean water: Surface Seawater, Deep Ocean Water, and Very Deep Ocean Water.

Each layer remains separate and autonomous from the others, moving at different speeds and in different directions due to different kinetic forces and different temperatures, densities, and life forms.

- *First: The surface seawater layer* is influenced by sunlight penetration and circulates rapidly in unison with the seasons and wind patterns to a depth of 250 m. It supports micro and animal life.

- *Second: The middle layer is DOW* where the water is free of sunlight and life forms. It is characterized not only by its mineral density but cold temperature, cleanliness, and trace elements. DOW is present at depths of between 250 and 1500 m. The deep ocean current moves very slowly under the influence of density and temperature gradients. The high mineral density is attributed to the depth-related pressure and the change in temperature from 20°C+ at the surface to 8°C at 600 m depth, which generates the movement of this layer.

- *Third: Very deep ocean water* has been discovered in a number of troughs in the Atlantic and Pacific Oceans. Depths can range from 1500 m to 15 km and life forms are supported where volcanic processes bring heat and minerals to the seabed floor.

DOW (Deep Ocean Water) creation begins when the summer ice melts from both Greenland and the subarctic region. The melting water collects minerals and trace elements during its journey to the ocean.

The minerals make the water heavier (DOW), so the water naturally sinks to the ocean floor where it commences a 2000-year-long journey. It flows southwards down the Atlantic Ocean, moves around the African Cape and then inches north through the Indian Ocean and into the western Pacific Ocean, first coming close to land at Taiwan, then Okinawa and Hawaii, and then arching back south toward Antarctica, where the changing seawater temperatures from the summer sun force the deep ocean water to the surface to feed the largest micro and macro food chain on our planet.

Over the past 15 years, there have been many new publications (over 40) establishing DOW as statistically significant with regards improved cardiovascular and metabolic function. Recent clinical research from Taiwan, Japan, and Korea also shows statistically.

3.5.2 Deep Sea Resources

The deep sea contains many different resources available for extraction, including silver, gold, copper, manganese, cobalt, and zinc. These raw materials are found in various forms on the sea floor, usually in higher concentrations than terrestrial mines. This is illustrated in Table 3.4.

3.5.3 Evaporating Seawater

Historically, the significance of the evaporative concentration of marine brines as indicators of the chemical evolution of seawater with time is reported in the literature, as shown in Figure 3.4. The precipitation of salts

TABLE 3.4

Raw Materials Found in Sea Floor

Type of Mineral Deposit	Average Depth	Resource Found
Polymetallic nodules	4,000–6,000 m	Nickel, copper, cobalt, and manganese
Manganese crusts	800–2,400 m	Mainly cobalt, some vanadium, molybdenum, and platinum
Sulfide deposits	1,400–3,700	Copper, lead, and zinc, some gold and silver

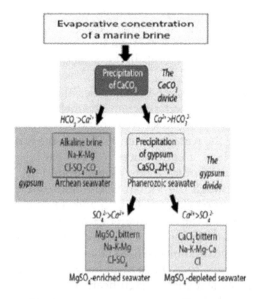

FIGURE 3.4
The evolution of three major marine brine types across deep time.

from evaporating seawater was thoroughly presented, as shown in Figure 3.5 where the composition changes in the following order:

First, calcite precipitates, followed by gypsum, halite, and various chlorides and sulfates. When it comes to the extraction of magnesium chloride from seawater as shown in Figure 3.6, the whole sequence of different salt separation is identified including $MgCl_2$. The concentration factor for the

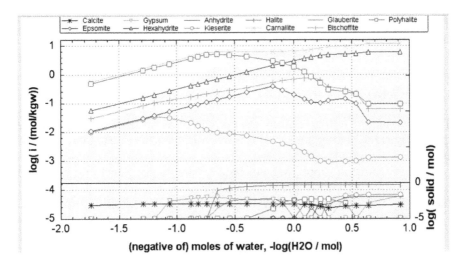

FIGURE 3.5
Evaporating seawater and precipitating salts.

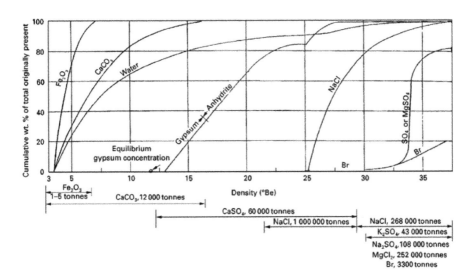

FIGURE 3.6
Deposition of salts during seawater evaporation.

different salts can be deduced from the cumulative amount of water evaporated, which is also given in the figure.

3.6 Economic Feasibility of Mineral Salts Recovery

Two factors determine the economic potentiality of extracting a mineral salt from seawater:

- The concentration of the mineral to be targeted.
- The value or the market price.

These two factors will determine the domain of the mineral as shown in the graph, Figure 3.7; either a mineral is economically attractive to be recovered, or not.

As seen from the graph, Mg enjoys a favorable position to be extracted commercially. The cost of production of primary magnesium is influenced mainly by the source of energy used and by the method of production.

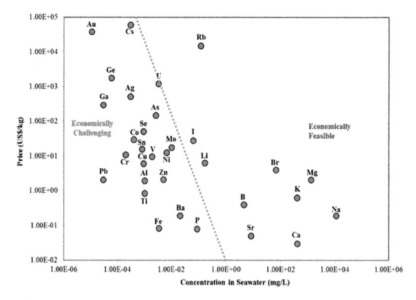

FIGURE 3.7

Classification of minerals based on concentration and values. (From Loganathan, P. et al., *Environ Sci: Water Res Technol*, 3, 37–53, 2017.)

China's domination of the magnesium market is a typical example. The Pidgeon process that depends on using dolomite as the main ore, ferro-silicon is the reducing agent, and coal as the energy source represents the main production method in China. It involves lower capital costs, but it is less environment-friendly relative to the electrolytic processes favored in the west. Details are covered in the text in latter chapters.

4

Production of Magnesium Chloride: An Overview

4.1 Introduction

Magnesium chloride is recognized as the strategic compound for the manufacture of magnesium metal. The principal precursor to magnesium metal is anhydrous magnesium chloride, which makes the potential feed to the electrolytic process. There are two main steps involved in the electrolytic process for the manufacture of magnesium metal: the preparation of a feedstock containing magnesium chloride, and the dissociation of this compound into magnesium metal and chlorine gas.

In addition to seawater, brine, and bitterns, magnesium chloride could be produced by dissolving magnesite ores in HCl, as shown in Figure 4.1.

Hydrated magnesium chloride is prescribed as oral magnesium supplements. There are three sources to recover magnesium chloride from:

- Naturally occurring brines, such as the Great Salt Lake (typically containing 1.1% by weight magnesium).
- The Dead Sea in the Jordan Valley, (3.4%).
- Oceans of the world (the largest source).

Although seawater is only approximately 0.13% magnesium, it represents an almost inexhaustible source. The hydrated magnesium chloride can be extracted from brine or seawater.

When it comes to magnesium ores, magnetite and dolomite, with a theoretical magnesium content of 47.6% and 22%, respectively, comprise the largest mineral sources of magnesium and magnesium compounds. In addition, sources of magnesium and its compounds include seawater, brines, and bitterns. Brine is very different when it comes to its concentration of salts. It has a high concentration of dissolved minerals, which could reach more than 50%, including magnesium. This makes brine an attractive route to recover magnesium as compared to seawater, which has around 3.5%

Three Sources to obtain Magnesium Chloride

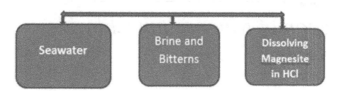

FIGURE 4.1
Sources of $MgCl_2$.

dissolved salts. Therefore, to make seawater production comparable to brine wells, a very large amount of seawater must be processed.

We find that the production of anhydrous magnesium chloride is the cornerstone in the process. At the same time, this is still the cause of most technological problems in the process.

The production of extracting magnesium chloride from seawater is detailed in this chapter.

4.2 Magnesium Chloride from Seawater

4.2.1 Dow Process

The Mg^{2+} cation is ranked second after Na^+. It could be classified as an abundant cation in seawater. This, by far, makes seawater and sea salt commercial sources for Mg.

The well-known Dow process represents a commercial source for the extraction of magnesium chloride from seawater. In brief, calcined dolomite is used in the precipitation of magnesium hydroxide from seawater. The calcium hydroxide reacts with the magnesium ions in seawater in a double displacement reaction. It is then separated and converted to $MgCl_2$ by reacting with hydrochloric acid. The chemical reactions are represented as follows:

$$MgCl_2 + Ca(OH)_2 \rightarrow Mg(OH)_2 + CaCl_2$$

$$Mg(OH)_2 + 2HCl \rightarrow MgCl_2 + 2H_2O$$

The hydrated magnesium chloride, which is the product obtained, as shown earlier, undergoes a drying process to bring the water content in the product to 1.5 moles of water. This makes it an anhydrous product to be introduced directly into the electrolytic cells. The by-product chlorine is produced from the electrolytic cells and is utilized by recycling it to manufacture hydrochloric acid, to be used as we will explain next.

Magnesium chloride is a compound that has six moles of water of hydration, written as $MgCl_2 \cdot 6H_2O$; that is attached to the polar Mg-Cl molecule as a result of intermolecular attractions. Therefore, the water molecules cannot be removed easily due to intermolecular attractions exhibited by this structure.

As stated earlier, simple drying techniques using heat remove water until the mixture reaches about 72% magnesium chloride and its structure is represented as $(MgCl_2 \cdot 1.5H_2O)$. However, further attempts to apply heat are not fruitful in this case. On the contrary, heat causes the remaining mixture to change to magnesium oxychloride. This is a stable refractory material.

In another version of Dow process, partly dehydrated magnesium chloride can be obtained by a different approach. Seawater is mixed in a flocculator with lightly-burned reactive dolomite. A reaction takes place forming an insoluble magnesium hydroxide, which precipitates, in the form of a slurry, to the bottom of a settling tank. The slurry is pumped from the tank to be filtered. To convert the hydroxide to the chloride, hydrochloric acid is used to bring in the magnesium chloride product. Drying is carried out by using a series of evaporation steps. This brings the water content in this product to 25%. Final dehydration takes place during smelting.

4.2.2 Extraction from Dolomite and Magnesite Ores

Magnesium hydroxide is produced as a result of the following consecutive steps: dolomite is crushed, roasted, and mixed with seawater in large tanks. Magnesium hydroxide settles to the bottom. By heating the magnesium hydroxide, followed by mixing with coke along with the presence of chlorine, a chemical reaction takes place, producing molten magnesium chloride. Once this product is obtained, the next step to produce magnesium is by the electrolysis. Magnesium is collected as it floats to the surface.

4.3 $MgCl_2$ Extraction from Desalination Plants

Desalination plants are recognized as vehicles designed mainly to give us fresh water from seawater, along with bottom product, called brine. This saline effluent from desalination plants is rich in *magnesium*. However, often this by-product is considered a *waste product*. The actual potentiality of the desalination processes has been presented in detail in a number of texts and will not be discussed here. However, cogeneration, or *dual purpose* plants, namely by definition, are those that use a single energy source to perform several functions. When this concept is applied in desalination, it has the obvious application to improve integration and efficiency of water and salt production.

Following this concept, a magnesium-chloride rich solution has several important uses as a raw material: to produce magnesium oxide and magnesium metal. There are a few cases that consider this approach to processing brines from desalination plants a promising source to produce magnesium. With the technology available today, it can change the paradigm of *waste product* into a *resource* aiming for the utilization of brine to take place.

The basic phenomena for extraction $MgCl_2$ from seawater is based on the physical and chemical natural phenomena occurring when seawater evaporates. In addition to pure distilled water (the main product of the cogeneration plant, used as potable water), seawater evaporation products include NaCl, MgCl, and other salts that contain impurities of calcium sulfate. Among the dissolved ions that we find in seawater, NaCl is, by far, the most abundant, corresponding to 85.65% of all the dissolved salts (mass).

4.4 Dual-Purpose Solvay-Dow: Conceptual Process

Abdel-Aal and coworkers (2016) proposed a single process for the manufacture of soda ash and magnesium chloride. It does so by first subjecting salt brines to ammonia, causing two simultaneous actions:

1. The absorption of ammonia, forming what is called *ammoniated brine*.

2. The precipitation of magnesium ions, found in the brine as magnesium hydroxide, $Mg(OH)_2$, which is filtered and separated. Next, carbon dioxide is introduced through the bulk of ammoniated brine (brine saturated with ammonia), causing the chemical conversion of both Na^+ and Cl^- into $NaHCO_3$ and NH_4Cl, respectively. Soda ash (Na_2CO_3) and ammonium chloride NH_4Cl come as products, along with partially desalinated water. This is detailed in the next section.

4.5 Modified Desalination Scheme

4.5.1 Introduction

Technology possesses immense potential to aid in solving the world's most pressing problems. Through research, it is concluded that one of today's most needed breakthroughs is a cost-effective, energy-efficient method for desalinating water using renewable energy. A breakthrough desalination technology could mitigate future problems associated with the need of a water supply.

Along with the existing trends in the chemical process industries, it is found that a strong bias exists toward integrated processing, cogeneration, and minimization of waste product generation, Abdel-Aal (2014). This gave the lead to initiate work toward our current proposal through what may be called coprocessing and coproduction. Considering the necessity to address theoretical and experimental investigations for an alternative route, this work presents a preliminary technical analysis for a sustainable, modified, multipurpose desalination scheme. The challenge of further improving desalination schemes is addressed through the management by coprocessing of reject brine.

Reject brine exiting desalination plants is a highly-concentrated waste by-product. It is estimated that for every 1 cubic meter of desalinated water, an equivalent amount is generated as reject brine. The common practice in dealing with brine is to discharge it back into the sea, where it could result, in the long run, in having detrimental effects on the aquatic life as well as the quality of the seawater available for desalination in the area.

4.5.2 Concentrate Management

The lack of economically and ecologically feasible concentrate management options is a major barrier to the widespread implementation of desalination. The main concern in desalination is the management of the brines, whose uncontrolled discharge has significant negative impacts on the environment. Options of concentrate management are shown next:

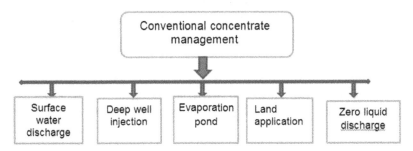

To these conventional methods, coproduction or coprocessing represents a promising alternative.

4.5.3 Objective

Science is about more than just what happens in the lab. There are other factors, including economic forces, affecting the kinds of questions we ask and how we go about trying to answer them. This is typically true for desalination; in particular, how should we handle reject brine and recover valuable products, such as magnesium chloride?

Our present work focuses on utilizing these reject brines as a feedstock for further processing. This way, we are not only managing a solution to handle reject brine, but we are creating a means for additional income to the water desalination industry. Our proposal is a kind of a research into a suite of seawater desalination improvements in order to make the process cheaper and more environmentally friendly. Two main reasons in support for this proposed scheme are:

1. It eliminates environmental problems, from displacing ocean-dwelling creatures to adversely altering the salt concentrations around them. No reject brine is dumped back to the sea.

2. It provides additional income from the sales of valuable products by the integration of the chemical conversion unit (CCU) that recovers magnesium chloride in addition to other chemicals.

4.5.4 The Road Map to Our Proposed Multipurpose Desalination Scheme

Our proposed overall desalination scheme is represented first by the following black box shown next:

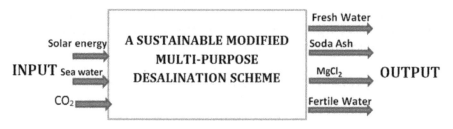

Now, to reveal the contents of this black box, our proposed scheme consists basically of three main units: MSF, brine concentration unit, and chemical conversion unit (CCU) are shown next:

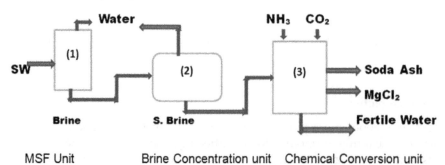

In addition, an auxiliary unit is hooked into the set-up to supply a solar heat source for the whole plant and to provide the required raw materials for unit 3.

4.5.5 Solar Evaporation Unit (Brine Concentration)

Brine exit desalination unit has to be concentrated from about 6% sodium chloride to saturated brine, about 26%–30%. This is a prerequisite for the next stage that takes place in the desalting unit. Evaporation is carried out using heat provided by the solar flux, as we will explain next. This unit provides additional fresh water to our system.

4.5.6 Chemical Conversion Unit

The chemical conversion unit (CCU) uses saturated brine as feed. Chemical reactions taking place will lead to the production of soda ash, magnesium chloride, and fertile water that contains ammonium chloride.

4.5.7 Auxiliary Units

The unit consists basically of three sub-units:

1. Source of solar heat supply for the whole desalination scheme. It is proposed to be a parabolic solar collector, capable of providing heat to the desalination unit, water electrolysis for hydrogen production, and more.
2. Renewable hydrogen production by water electrolysis using solar energy.
3. Ammonia synthesis.

Solar energy is a cornerstone in our proposal. The facilities provided by the auxiliary unit are exemplified in Figures 4.2 and 4.3.

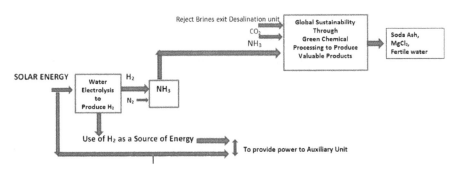

FIGURE 4.2
Role of solar energy to provide hydrogen for ammonia synthesis by water electrolysis.

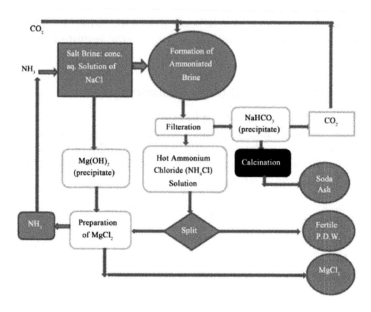

FIGURE 4.3
The modified Solvay-Dow process for magnesium. (From Abdel-Aal, H.K. et al., *Open Acce. Lib. J.*, 3, e2998, 2016.)

4.5.8 Case Study: Process Analysis and Material Balance Calculations

In order to carry out this study, the following two-step procedure is outlined:

First: Define the chemical reactions underlying our system. They are as presented next:

Main Reactions

The basic reactions involved could be visualized to take place as follows:

Reaction between CO_2 and NH_3 can be described as:

$$CO_2 + 2NH_3 \rightarrow NH_2COO^- + NH_4^+ \tag{4.1}$$

In the bulk of the solution, the carbamate hydrolyses comparatively slowly to bicarbonate:

$$NH_2COO^- + H_2O \rightarrow NH_3 + HOCOO^- \tag{4.2}$$

In the presence of NaCl, the following instantaneous reaction takes place:

$$NH_4^+ + HCO_3^- + NaCl \rightarrow NaHCO_3 + NH_4Cl \tag{4.3}$$

Soda ash is produced as an end product:

$$2NaHCO_3 \rightarrow Na_2CO_3 + CO_2 + H_2O \tag{4.4}$$

This leads to the precipitation of sodium bicarbonate leaving ammonium chloride in a partially desalinated water.

In addition, the following reaction could take place, leading to the production of $MgCl_2$, as well.

$$2NH_4Cl + Mg(OH)_2 \rightarrow MgCl_2 + 2NH_3 + 2H_2O \tag{4.5}$$

Second: Apply the stoichiometric principles to these reactions, as presented in the next table:

	Reactants					Products						
	NH_4	H_2O	CO_2	$MgCl_2$	$NaCl$	NH_4OH	$Mg(OH)_2$	$NaHCO_3$	Na_2CO_3	NH_4Cl	CO_2	Kg
No. 1	-4	-4				4						
No. 2			-1			-2	1			2		
No. 3		-2		-2	-2			2		2		
No. 4								-2	1		1	
Net	-4	-4	-2	-1	-2	0	1	0	1	4	1	

Now we are ready to seek the solution of our system, defined earlier as *chemical conversion unit* (CCU)—described earlier—by using *Excel* computations, as applied for the following case study:

Assuming that a feed of 7,700,000 gallon/day of seawater (equivalent to about 30,000 cubic meter/day), with a salinity is 35,000 ppm is to be desalinated using an MSF plant. For calculation purpose, this feed input is converted to mass to be 30×10^6 Kg/day, approximately, and is to be taken as a basis. The feed input, when flashed in the MSF plant, produces 15×10^6 Kg/d of brine with a salt concentration of about 7%. Assume also that for every volume of desalinated water, an equivalent amount is generated as reject brine.

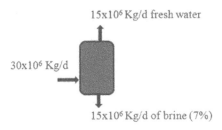

15×10^6 Kg/d fresh water

30×10^6 Kg/d

15×10^6 Kg/d of brine (7%)

The reject brine is then subjected to solar concentration/evaporation to raise the salt content in the brine to the saturation point, nearly 30% concentration. A material balance on this concentrator is carried out as follows:

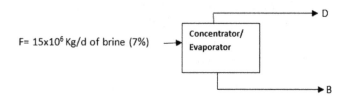

Total balance $F = D + B$

Component mass balance on the salt: $15 \times 10^6 (0.07) = B (0.3)$

Therefore, $B = 3.5 \times 10^6$ Kg/d of concentrated brine to be processed next for the chemical conversion step. The distillate rate of the fresh water obtained, as a result of this evaporation process, is $D = 11.5 \times 10^6$ Kg/d.

Again, taking one day as a basis in our next calculations, we simply drop per day.

Results of Excel Computations

	NaCl	NaHCO$_3$	NH$_4$Cl
M Wt	58.5	84	53.5
Feed water to MSF	3.00E+07	Kg	
Fresh water exit MSF	1.50E+07	Kg	
Brine 7% exit MSF	1.50E+07	Kg	
Brine 30% exit concentration	3.50E+06	Kg	
Fresh water exit concentration	1.15E+07	Kg	
NaCl in brine	1.05E+06	Kg	
Moles of NaCl	1.79E+07	Mole	
Moles of NaHCO$_3$	1.79E+07	Mole	
Moles of NH$_4$Cl	1.79E+07	Mole	
Moles of NH$_3$	1.79E+07	Mole	
Moles of CO$_2$	1.79E+07	Mole	
Moles of Na$_2$CO$_3$	8.97E+06	Mole	
Quantity of NaCl	1.05E+06	Kg	
Quantity of NaHCO$_3$	1.51E+06	Kg	
Quantity of NH$_4$Cl	9.60E+05	Kg	
Quantity of NH$_3$	3.05E+05	Kg	
Quantity of CO$_2$	7.90E+05	Kg	
Quantity of Na$_2$CO$_3$	9.51E+05	Kg	

- $NaHCO_3$ does not come as a final product. The final product we obtain is Na_2CO_3, known as soda ash, according to the reaction:

$$2NaHCO_3 \rightarrow Na_2CO_3 + CO_2 + H_2O$$

- This process is identified as *brine desalting* because of the drastic reduction of the salt content in brine.

Consumption Rate of Reactants versus Rate of Production of Products (Tons Per Cubic Meter of Reject Brine Exit MSF)

(Per Cu. Meter of Reject Brine Exit MSF)	Reactants	Products
NH_3	2.03E-02	
CO_2	5.26E-02	
NH_4Cl		6.40E-02
Na_2CO_3		6.34E-02

This work presents a preliminary technical analysis for a sustainable, modified, multipurpose desalination scheme. The challenge of further improving desalination schemes is addressed through the management by coprocessing reject brine.

Along with the existing trends in chemical process industries, it is found that a strong bias exists toward integrated processing, cogeneration, and minimization of waste product generation, Abdel-Aal (2014). This gave the lead to initiate work towards our current proposal through what may be called coprocessing and coproduction.

Two main reasons are in support of this proposed scheme:

- The elimination of environmental problems, from displacing ocean-dwelling creatures to adversely altering the salt concentrations around them. No reject brine is dumped back to the sea.
- The integration of the chemical conversion unit (CCU), to be part of desalination plants will lead to additional income from the sales of valuable products.

The results of the *Excel* computations given previously reveal very interesting conclusions:

First: The dollar values of the products obtained from the proposed system should add to its merits. These include two main products: soda ash and ammonium chloride. The latter could be produced, dissolved in partially desalted water, to be sold as a fertilizer.

Second: The recovery of magnesium as magnesium chloride makes the third product to be recovered. It is a very viable option. This will promote the application of our proposed desalination system since magnesium chloride is the raw material to produce magnesium metal.

Third: When we say sustainable energy, we are basically referring to renewable energy resources, in particular, solar energy. Hence, a sustainable desalination system would be one that is provided by a source of solar heat supply for the whole desalination scheme. It should be capable to provide heat to the desalination unit, brine evaporation/concentration, and for water electrolysis for hydrogen production as well.

The concentration (evaporation) step of the reject brine exiting the desalination unit to bring it to about 27% sodium chloride is of prime importance. Parabolic solar collectors are recommended for this task.

Fourth: To have a reliable and cheap source of ammonia, it is suggested to produce the hydrogen for ammonia synthesis by water electrolysis using solar energy. Figure 4.2 is a schematic representation of this cycle. Solar energy is a cornerstone in our proposal.

Fifth: The option of producing fertile water (partially desalted water) containing NH_4Cl to be used for agriculture purposes is a good choice. Salt content in this water is reduced from initial brine concentration of 25% to about 7%. Ammonium chloride is an excellent fertilizer used in the Far East for rice crops, and it is recommended as an extremely good source of both nitrogen and chloride for coconut oil.

Sixth: The proposed desalination scheme should find widespread support in many of the Arab Gulf countries such as Saudi Arabia and the United Arab Emirates. They will produce fresh water to drink, along with valuable chemical products to sell, without harming the marine environment by reject brines exit desalination plants.

Seventh: This process is identified as *brine desalting* because of the drastic reduction of the salt content in brine.

Magnesium chloride ($MgCl_2$) is obtained, next, by reacting ammonium chloride (NH_4Cl) with $Mg(OH)_2$. In this double reaction, ammonia will be regenerated and recycled back to the process:

$$2NH_4 + Mg(OH)_2 \rightarrow MgCl_2 + 2NH_3 + 2H_2O$$

The presented contribution offers this conceptual scheme as an amalgamation of both Solvay/Dow (Magnesium) process. The *modified* Solvay process is shown in Figure 4.3.

4.6 Seawater Bitterns

Electrolytic processes worldwide are the working-horse of most of the large magnesium production plants. These processes generally require magnesium chloride as a feed to be electrolyzed by an electric current to convert it to magnesium metal and chlorine gas. In some cases, brine or evaporated bitterns are used as a feed source to produce magnesium chloride.

Seawater bitterns (SWB) and brines represent a good source for magnesium manufacture (Abdel-Aal and coworkers, 2017). SWB are encountered in the processes of desalination and sea salt production where large quantities of bitterns and brines are produced as by-products. The term *bitterns* refers to the very bitter-tasting solution that remains after evaporation and crystallization of sodium chloride (table salt) from brines and seawater.

It is a concentrated form of a collection of magnesium, potassium, sulfate, and chloride salts, such as KCl, $MgCl_2$, $MgSO_4$, and double salts. SWB could be described as *exhausted brines*. Brine, on the other hand, designates a solution of salt (usually sodium chloride) in water. In a different context, brine may refer to salt solutions ranging from about 3.5% (a typical concentration of seawater, or the lower end of solutions used for brining foods) up to about 26% (a typical saturated solution, depending on temperature).

Other levels of concentration of salt in water are identified and given different names as shown next:

Fresh Water	Brackish Water	Saline Water	Brine
<0.05%	0.05%–3%	3%–5%	>5%

In salt-works, brines can represent a rich and promising source of raw materials, especially when they are very concentrated. In particular, magnesium concentration can reach values up to 30–40 kg/m³ of brine, which is 20–30 times that of typical seawater. In theory, for every ton of sea salt produced, about 1 m³ of bittern is produced. Chemicals found in the bittern corresponding to 10 million tons of salt produced are classified into the different types in Table 4.1.

It is conclusive from the earlier that bitterns are rich in chemicals and can be described as a *precursor* for the production of magnesium metal. Brines,

TABLE 4.1

Chemicals Produced from Bitterns

Salt Produced	Tons
NaCl	1,500,000
$MgCl_2$	1,200,000
$MgSO_4$	700,000
KCl	238,000
Bromine	20,000

on the other hand, are important sources of salt, iodine, lithium, magnesium, potassium, bromine, and other materials, and are potentially important sources of a number of others. As we have seen earlier, the brines may be seawater, other surface water, or groundwater.

5

Production of Magnesium Chloride from Seawater: Proposed Method-Preferential Salt Separation

5.1 Introduction

To exploit valuable salt products, in particular $MgCl_2$, from seawater in desalination plants and/or salt production, various experiments were carried out, in particular by the author and his colleagues. Mainly they consisted of applying the physical concept of preferential type of separation, with the understanding that $MgCl_2$ is the most soluble salt that will separate at the very end.

Kettani and Abdel-Aal (1973) proposed a physical separation method known as the Preferential Salt Separation (PSS) to obtain magnesium chloride directly from seawater. In principle, the PSS concept is based on the selective separation of salts during the evaporation process of a dynamic flow of a fluid of brines.

5.2 Preferential Salt Separation

5.2.1 Background

Water is a very good solvent. Solvents are liquids that dissolve other substances. Most of the water on earth, including the water in oceans, lakes, rivers, and ponds, contains many solutes.

Soluble salts dissolved in seawater crystalize out of the solution at different concentration levels. When brine is concentrated by evaporation and loses its water content, it leads to the successive precipitation of the following salts in this order: the least soluble salts, $CaCO_3$ and $CaSO_4$, come first, followed by the separation of NaCl and finally Mg and K.

The work published by *manoa.hawaii.edu* and shown in Figure 3.2 identifies the salt rings formed when seawater was evaporated from a watch glass. The outer ring, which precipitated out of solution first, is primarily made up of calcium carbonate ($CaCO_3$). Carbonates are the least soluble salts in seawater. The inner ring is primarily made up of potassium (KCl) and magnesium ($MgCl_2$) salts, which are very soluble.

This type of experiment, which was carried out using watch glass, is very significant and informative for our current presentation on the Preferential Salt Separation (PSS) process for the extraction of magnesium from seawater. This experiment is supportive of the basic principle behind the PSS; that is magnesium chloride is the last salt to precipitate from evaporating seawater.

To exploit valuable salt products as by-products, few methods are currently available; magnesium chloride ($MgCl_2$), that is, the main feedstock for the electrolysis phase to produce magnesium metal, has been traditionally produced from seawater by precipitating it as magnesium hydroxide, then converting it to the chloride by adding hydrochloric acid. The PSS method, on the other hand, is different in concept, since it is totally a physical operation.

5.2.2 Experimental Work

The PSS model was experimentally tested for the first time on a semi-pilot scale. The evaporator consists of a number of shallow inclines (channels), arranged in a step-wise pattern with the dimensions 2 m by 3 m. The channel is 10 cm wide. Pre-concentrated saline water (brine) is continually fed in a dynamic flow to the upper basin, then allowed to overflow by gravity to subsequent channels all the way to the very end.

Heat flux necessary for the evaporation of the water (brine) was provided by a heating system that used infrared heaters, as shown in Figure 5.1.

A development took place later on by heating the evaporator using solar energy. Heat flux necessary for the evaporation of the water is provided by the Joule Ohmic heating system with the subsequent selective separation of salts along the evaporator. This takes place according to the solubility product of each mineral salt.

Water condensing on the inside of the plastic cover is to be collected in a trough installed on each basin. Flow conditions and the level of saline water in the basins are controlled to resume the optimum performance of operation. Whatever the evaporative process might be, the necessary end product of the evaporative sequence would be a dense magnesium solution.

An industrial model based on this concept was proposed to be built and located near the seashore. It consisted of a number of shallow inclined basins (10 m wide by 10 m long), arranged in a step-wise pattern. In each

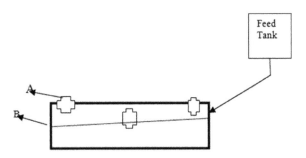

FIGURE 5.1
Preliminary experimental setup for PSS unit. Legend: (a) infrared light heaters and (b) inclined evaporators.

basin covered with a transparent plastic cover, pre-concentrated saline continually fed to the upper basin, and was allowed to overflow by gravity to the next channels all the way to the very end. Solar energy, absorbed by the water and by the black polyethylene film covering the bottom of the evaporator, would supply the heat flux necessary for evaporation with the subsequent selective separation of salts along the evaporator, according to the solubility product of each salt. Water condensing on the inside of the plastic cover would have been collected in a trough installed on each basin. Whatever the evaporative process might be, the necessary end product of the evaporative sequence would be a dense magnesium solution. The concentrated solution is then dried (dehydrated to yield granular of anhydrous magnesium chloride of approximate composition $MgCl_2 \cdot 1.5H_2O$, which contains about 74% $MgCl_2$).

5.2.3 Additional Experimental Work

Work was further extended by running a simulation program for Al-Khobar Desalination Plant in Saudi Arabia. Real data was used through 32 repeated cycles of 4 stages each. Additional potable water, NaCl, and a highly concentrated $MgCl_2$ bittern were obtained from the brines by Abdel-Aal et al. (1990). A theoretical study on the economics of a new magnesium production-process, carried out by Abdel-Aal (1982), stimulated interest in the subject. Experimental work was resumed to build a multipurpose solar desalination pilot unit (MPSDU) for the enhanced evaporation of saline water.

Another experiment was carried out for the selective separation of $MgCl_2$ bittern from seawater using a batch model. This model of evaporation utilizes the same principle of fractional separation but uses a batch operation.

5.3 Experimental Results Using Preferential Salt Separation—"Dynamic-Flow"

The progress of the separation of salts along the channel evaporator that consists of 20 channels is given in Figure 5.2.

The separation sequence of these salts is best described by assigning the concentration of the salts in ppm. The progress of separation of magnesium is described as follows:

Its concentration increases steadily from 1,850 ppm at the inlet up to 14,400 ppm, corresponding to a sp. gravity = 1.215 (seawater feed). With further evaporation along the evaporator, it drops to 10,000 ppm; this signals the precipitation of $MgSO_4 \cdot nH_2O$. With further progress through more channels, Mg concentration shoots to a peak of about 29,000 ppm, indicating that $MgSO_4$ precipitation is coming to an end. The peak itself shows that the new salt (bittern) consisting partly of Mg is reaching maximum saturation (probably carnalite [$KCl \cdot MgCl_2 \cdot 6H_2O$]); hence, precipitation occurs once saturation is reached. On channel 17, precipitation reaches its maximum, corresponding to Mg concentration of 14,000 ppm in the remaining brine. Further evaporation leads to a sharp increase in Mg concentration to a value of 42,000 ppm almost at the end of the evaporator. This indicates that the brine is highly concentrated in Mg, which can be described as a bittern or magnesium-rich brine. Comparing the ratio of Mg

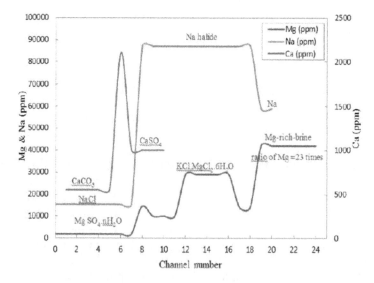

FIGURE 5.2
Progress of separation of salts along the PSS evaporator.

concentration in the final output obtained to that in the intake seawater, it is found to be $= 42,000/1,850 \approx 23$ times higher concentration.

5.4 Preferential Salt Separation—"Static-Flow" or Batch Model

This model of evaporation utilizes the same principle of fractional separation but by using a batch operation. The evaporation takes place in a tank using a solar-heated water jacket. When salts start to precipitate, evaporation is stopped and salts are separated by filtration. Chemical analysis is carried out to mark the first sample of salts. Evaporation is resumed, and the procedure is repeated for additional salt precipitates until the stage of magnesium-rich brine (bittern) is reached.

5.4.1 Experimental Results

The batch model was implemented as described earlier. A preliminary experiment was carried out on a lab scale using a sample of 2 L of seawater. Results are shown in Table 5.1. It is evident that a similar trend in salt separation is followed in this model as well. Magnesium-chloride-rich brine is observed in terminal samples. Because of the limited scale of the initial volume used, very small quantities of separated salts were obtained.

5.5 Proposed Small-Scale Integrated Pilot Model

Since the primary target is to produce magnesium metal, the overall scheme of such a project could be simply represented, which is shown next:

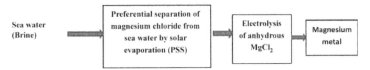

For an industrial implementation, it will be assumed that solar energy will be fully utilized to power the PSS system. Taking an output of 2 kg per day of magnesium chloride, as a basis, the system is to be powered by a 2000 m^2 photovoltaic roof in order to provide power for the PSS unit (evaporation and concentration) to produce anhydrous magnesium chloride and to run the electrolytic cells for magnesium production as well.

TABLE 5.1

Experimental Results for Batch Model

Original Water: Seawater		PPT 1	PPT 2	PPT 3	PPT 4	PPT 5	PPT 6	PPT 7
PH	8.42					7.1	7.1	6.5
TDS	35 gm/L					5.423	5.498	4.865
Conductivity	57.5 ms/cm							
Density gm/mL	1.0264	1.04436	1.22125	1.22705	1.22905	1.21562	1.2122	1.217
	Al, mg/L	<0.01	<0.01	<0.01	<0.01			
	Ba, mg/L	0.0046	0.0103	0.0058	0.0031			
	Ca, mg/L	106.4	47.25	14.17	8.613	41.8	8.8	19.8
	Cd, mg/L	<0.001	<0.001	<0.001	<0.001			
	Co, mg/L	<0.0004	<0.0004	0.0005	0.0005			
	Cr, mg/L	<0.003	0.0045	<0.003	<0.003			
	Cu, mg/L	<0.003	0.0045	<0.003	<0.003			
	Fe, mg/L	<0.02	<0.02	<0.02	<0.02			
	Mg, mg/L	0.4536	24.76	21.38	5.198	2.41	0.8019	7.22
	Mn, mg/L	<0.001	<0.001	<0.001	<0.001			
	Mo, mg/L	0.0013	<0.0005	<0.0005	0.0046			
	Ni, mg/L	<0.001	<0.001	<0.001	<0.001			
	Pb, mg/L	0.0093	<0.004	<0.004	<0.004			
	Sr, mg/L	3.027	0.8726	0.2141	0.2146			
	V, mg/L	0.0106	0.0205	0.0114	0.013			
	Zn, mg/L	<0.001	0.0064	0.008	<0.001			
	K, mg/L	5	24	32	38	3	2	3
	Na, mg/L	32	110	1420	1480	1900	2100	1900

5.6 Mathematical Model for the Evaporation of Seawater along the Preferential Salt Separation: Flow of a Falling Film across an Inclined Flat Surface

5.6.1 Introduction

The basic idea of the PSS, as described before, is to allow for the successive salt separation of the flowing brine along an inclined evaporator. In this section, a mathematical model is presented to simulate the evaporation process across the channels of the evaporator.

Evaporation is conducted for different cases as described next.

First: Using the rate of solar evaporation of water under direct conditions. It is estimated to be about 3×10^{-4} mm/sec. This is based on 3.5 m/year as an average.

Second: Using *enhanced* solar evaporation by heating for the following cases.

5.6.2 Theoretical Background

This is the case of a flow of fluids along an inclined flat surface, as presented in Figure 5.3.

According to Bird and Associates, such films have been studied in connection with wetted wall columns and evaporation and absorption experiments.

Using this model, it has been possible to calculate the velocity distribution (v) and the depth or the level of the fluid attained on these channels (H).

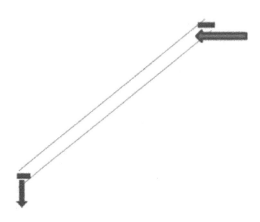

FIGURE 5.3
Inclined flat surface for a falling film.

The basic equations used are:

- The average velocity,

$$v = \frac{\{\rho\, g\, H^2 \cos \varphi\}}{3\,\mu} \,..... \text{cm/sec}$$

- Volumetric flow rate,

$$Q = \frac{\{\rho\, g\, H^3\, W \cos \varphi\}}{3\,\mu} ... \text{cm}^3/\text{sec}$$

- Residence time,

$$\theta = \frac{w}{v} \,..................... \text{sec}$$

where:
  ρ and μ are the density and viscosity of the liquid $= 1.3 \text{ gm/cm}^3$ and 4×10^{-2} (gm/cm.sec), respectively
  g is the gravitational force, 980 cm/sec
  φ is the inclination of the basin with the vertical direction, for the PSS evaporator $= 85°$ cos. $\varphi = 0.1736$
  w and L are the width and length of the basin $= 1.0$ and 0.5 m
  H is the Height of fluid in the basin $= 2$ cm initially, at the start of the evaporation

The solution to calculate the number of channels needed for evaporation using PSS is illustrated in Figure 5.4. The calculation is done for two cases:

First: Using rate of solar evaporation of water under direct conditions. It is estimated to be about 3×10^{-4} mm/sec. This is based on 3.5 m/year as an average.

Second: Using *enhanced* solar evaporation.

5.6.3 Results

First: Using rate of solar evaporation of water under direct conditions. It is estimated to be about 3×10^{-4} mm/sec. This is based on 3.5 m/year as an average.

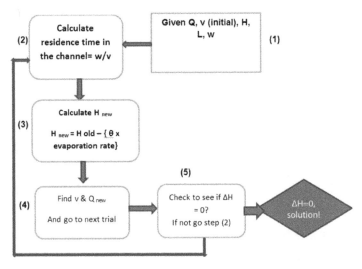

FIGURE 5.4
Outline for the procedure to calculate the number of channels for PSS. Where ΔH = [H for trial n] – [H for trial n + 1]. That is, H decreases by evaporation from channel to the next channel.

To obtain magnesium chloride as end product using the PSS. it is evident from the data given in Table 5.1 that the evaporation rate is very slow and requires a big number of channels.

Second: Using *enhanced* solar evaporation.

Enhanced evaporation was tested for different solar inputs, as shown in the data obtained next (Table 5.2). The height of input fluid in channel 1 was drastically reduced by enhanced solar evaporation from 2 to 1.1378 cm using a solar input equivalent to 6 mm/sec (case 5). Data for PSS calculations are given in Table 5.3. However, trial calculations to find the number of stages using this approach are not included due to space limitations. (Table 5.3).

TABLE 5.2

Sample Calculations Using Direct Solar Evaporation

Channel	H (mm)	Q (cm³/sec).10⁶
1	20	1.4860
2	19.998	1.4858
3	19.996	1.4854
4	19.994	1.4852
5	19.992	1.4840

TABLE 5.3

Data for PSS Calculations

Microsoft Excel non-commercial use - Mathematical Model for the Evaporation of Seawater along the PSS.xlsx [Read-Only]

		B	C	D
1				
2		5.6 Mathematical Model for the Evaporation of Seawater along the PSS: Application of the		
3		Equation of Change for Design Calculation for the PSS		
4		case 1 : Rate of evaporation =		0.003 mm/sec
5		case 2 : Rate of evaporation =		0.3 mm/sec
6		case 3 : Rate of evaporation =		2 mm/sec
7		case 4 : Rate of evaporation =		4 mm/sec
8		case 5 : Rate of evaporation =		6 mm/sec
9		case 6 : Rate of evaporation =		7 mm/sec
10		ρ =	1.3	gm/cm3
11		g =	980	cm/sec2
12		Cos ϕ =	0.087155	
13		μ =	0.04	[gm/cm.sec]
14	W	width of the basin =	200	cm
15	L	length of the basin =	500	cm
16	H	H = Height of fluid in the basin =	2	cm
17	v	The average velocity, v = {ρ g H2 Cos ϕ }/ 3μ	3701	cm/sec
18	Q	Volumetric flow rate, Q = {ρ g H3 W Cos ϕ }/3 μ	1480473	cm^3/sec
19	θ	Residence time, θ w/v =	0.0540368	sec

Microsoft Excel non-commercial use - Mathematical Model for the Evaporation of Seawater along the PSS.xlsx [Read-Only]

		B	C	D	E	F	G
18	Q	Volumetric flow rate, Q = {ρ g H3 W Cos ϕ }/3μ	1480473	cm^3/sec			
19	θ	Residence time, θ w/v =	0.0540368	sec			
20		Rate of solar evaporation of water	3.00E-04	mm/sec	3.5 m/year as average		
21	H$_{new1}$	H$_{new}$ at Rate of solar evaporation of water (case 1)	1.99983789	cm	H1=H-(0*evaporation rate)		
22							
23	v2	The average velocity, v = {ρ g H2 Cos ϕ }/ 3μ	3700.5824	cm/sec			
24	Q2	Volumetric flow rate, Q = {ρ g H3 W Cos ϕ }/3 μ	1480113	cm^3/sec			
25	θ2	Residence time, θ w/v =	0.0540455	sec			
26	H$_{new2}$	H$_{new}$ at Rate of solar evaporation of water (case 2)	1.983624225	cm	H2=H1-(H1*evaporation rate)		
27							
28	v3	The average velocity, v = {ρ g H2 Cos ϕ }/ 3μ	3640.8207	cm/sec			
29	Q3	Volumetric flow rate, Q = {ρ g H3 W Cos ϕ }/3 μ	1444404	cm^3/sec			
30	θ3	Residence time, θ w/v =	0.0549327	sec			
31	H$_{new3}$	H$_{new}$ at Rate of solar evaporation of water (case 3)	1.873758888	cm	H3=H2-(θ3*evaporation rate)		
32							
33	v4	The average velocity, v = {ρ g H2 Cos ϕ }/ 3μ	3248.6872	cm/sec			
34	Q4	Volumetric flow rate, Q = {ρ g H3 W Cos ϕ }/3 μ	1217451	cm^3/sec			

Microsoft Excel non-commercial use - Mathematical Model for the Evaporation of Seawater along the PSS.xlsx [Read-Only]

		B	C	D	E	F	G
30	θ3	Residence time, θ w/v =	0.0549327	sec			
31	H$_{new3}$	H$_{new}$ at Rate of solar evaporation of water (case 3)	1.873758888	cm	H3=H2-(θ3*evaporation rate)		
32							
33	v4	The average velocity, v = {ρ g H2 Cos ϕ }/ 3μ	3248.6872	cm/sec			
34	Q4	Volumetric flow rate, Q = {ρ g H3 W Cos ϕ }/3 μ	1217451	cm^3/sec			
35	θ4	Residence time, θ w/v =	0.0615633	sec			
36	H$_{new4}$	H$_{new}$ at Rate of solar evaporation of water (case 4)	1.627505572	cm	H4=H3-(θ4*evaporation rate)		
37							
38	v5	The average velocity, v = {ρ g H2 Cos ϕ }/ 3μ	2450.8992	cm/sec			
39	Q5	Volumetric flow rate, Q = {ρ g H3 W Cos ϕ }/3 μ	797770	cm^3/sec			
40	θ5	Residence time, θ w/v =	0.0816027	sec			
41	H$_{new5}$	H$_{new}$ at Rate of solar evaporation of water (case 5)	1.137889362	cm	H5=H4-(θ5*evaporation rate)		
42							
43	v6	The average velocity, v = {ρ g H2 Cos ϕ }/ 3μ	1198.0655	cm/sec			
44	Q6	Volumetric flow rate, Q = {ρ g H3 W Cos ϕ }/3 μ	272653	cm^3/sec			
45	θ6	Residence time, θ w/v =	0.1669358	sec			
46	H$_{new6}$	H$_{new}$ at Rate of solar evaporation of water (case 6)	-0.0306611	cm	H6=H5-(θ6*evaporation rate)		

5.7 Proposals for Other Separation Options: Separation of Magnesium Chloride from Sodium Chloride in Seawater by the Dense-Phase Technique

An experimental work was carried out by the author to investigate the possibility of separating a mixture of magnesium chloride and sodium chloride. It is described as follows.

Experimental results were reported for the solid–solid separation of a mixture of $MgCl_2 \cdot 6H_2O/NaCl$ (to be referred to as M/S), using a dense media of such organic solvents as tetrachloromethane, iodomethane, and a combination of these two solvents, was to be separated. The solid M/S mixture to be separated was prepared from two different sources: by evaporating seawater in which the M/S ratio is about 0.14, and by synthesis of the pure salts in which the M/S ratio varied between 0.5 and 1.275. Good separation results are reported for the case of dense-phase separation in which the blend of the two solvents has a specific gravity of 1.9. Recovery was about 95% pure $MgCl_2 \cdot 6H_2O$ from a synthetic mixture having an M/S ratio of 1.0.

6

Commercial Methods for Magnesium Production

6.1 Introduction/Classification

There are basically two commercial methods of producing magnesium: *electrolysis of molten magnesium* and *thermal reduction of magnesium oxide*. However, there is an additional method where magnesite is dissolved in HCl, forming MgCl. This method is applied in Australia. Classification is demonstrated next:

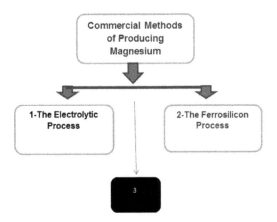

Comments:

1. Using electrolytic cells, the process involves the electrolysis of fused $MgCl_2$ in two consecutive steps: to make and prepare the anhydrous magnesium chloride first as a feedstock, and the dissociation of this compound into magnesium metal and chlorine gas second.

2. Using an iron-silicon alloy as a reducing agent, magnesium oxide (prepared by calcining dolomite) is reduced to magnesium metal.

3. Using hydrochloric acid to dissolve magnesite, magnesium chloride is produced. This technique is primarily used in Australia.

The magnesium production technique is determined by the location and the type of resource (feed). They are significant factors to consider. In addition, the fact that magnesium is abundant would promote its production in many locations. On the top of that, its end-use is very controlled by the product price. This will make the cost price a decisive factor in the magnesium market.

Magnesium is produced on a large scale worldwide using electrolytic processes. Basically, magnesium chloride is used in the process as a feed. It is electrolyzed using electric power, producing magnesium metal chlorine. Brine or evaporated bitterns are normally used as a feed in this electrolytic process to produce magnesium chloride.

Two steps are basically involved in the electrolytic process. The first step deals with the preparation of anhydrous magnesium chloride. The second step is the heart of the process in which dissociation of magnesium chloride takes place, producing magnesium and chlorine. This step is energy-intensive, consuming large amount of electric current.

The second method is known as the *ferrosilicon process*. In this method, magnesium oxide (prepared by calcining dolomite) is reduced to magnesium metal by using an iron-silicon alloy.

Table 6.1 presents a comprehensive list of different processing routes for magnesium production.

TABLE 6.1

Processing Options for the Production of Magnesium

Process Route	Sources	Feed Preparation	Reaction	Temperature/Pressure
Electrolytic				
Dow process[1]	Brine/seawater	Neutralization, purification, and dehydration	Electrolytic $MgCl_{2(l)} \rightarrow Mg_{(l)} + Cl_{2(g)}$ Cathode: $2Cl^- \rightarrow Cl_2 + 2e$	T = 700°C–800°C P = 1 atm
AM process[2]	Magnesite	Mining, leaching with HCl, and dehydration	Anode: $Mg^{2+} + 2e \rightarrow Mg$	
IG Farben process[1]	Seawater/brine	Neutralization, prilling, dehydration, and chlorination		
Thermal Reduction Process				
Silicothermic[3]	Dolomite FeSi	Calcination, FeSi making, and pelleting	$MgO + CaO + FeSi = Mg_{(g)} + Ca_2SiO_{4(s)} + Fe_{(s)}$	T = 1160°C P = 13–67 Pa $(1.2 \times 10^{-4} atm)$
Carbothermic[4]	Magnesite, carbon	Calcination and pelleting	$MgO + C = Mg_{(g)} + CO_{(g)}$	T = 1700°C P = 1 atm

(Continued)

TABLE 6.1 (*Continued*)

Processing Options for the Production of Magnesium

Process Route	Sources	Feed Preparation	Reaction	Temperature/ Pressure
Magnetherm[5]	Dolomite, bauxite, FeSi	Calcination and FeSi making	$2CaO \cdot MgO + (x\,Fe)$ $Si + n\,Al_2O_3 = 2CaO.$ $SiO_2.\ nAl_2O_3 + 2Mg + x\,Fe$	T = 1550°C P = 0.05 atm
Aluminothermic[6]	Dolomite Al scrap	Calcination	$4MgO_{(s)} + 2Al_{(s)} = 3Mg_{(g)} + MgAl_2O_{4(s)}$	T = 1700°C P = 0.85–1 atm
Mintek[7]	Dolomite, bauxite, FeSi, Al Scrap	Calcination	$2\ CaO \cdot MgO + (xFe)\ Si + n\,Al_2O_3 = 2CaO \cdot SiO_2.$ $n\,Al_2O_3 = 2Mg + x\,Fe$ $4MgO_{(s)} + 2Al_{(s)} = 3Mg_{(g)} + MgAl_2O_{4(s)}$	T = 1700°C P = 0.85 atm

Source: [1]Habashi, F., 'Magnesium': *Handbook of Extractive Metallurgy*, Wiley-VCH, Winheim, Germany, 1997; [2]Jenkins, D.H. et al., *Miner. Process. Extr. Metall. (Tran Inst. Min Metall. C)*, 118, 205–213, 2009; [3]Mayer, A., *Trans. AIME*, 159, 363–376, 1944; [4]Brooks, G. et al., *J. Miner. Metals Mater. Soc.*, 58, 51–55, 2006; [5]Faure, C. and Marchal, J., *J. Metals.*, 16, 721–723, 1964; [6]Wadsley, M.W., *Mag. Technol., Minerals, Metals, Mater. Soc.*, 2000, 65–70, 2000; [7]Schoukens, A.F.S. et al., *J. South Afr. Inst. Min. Metall.*, 106, 25–29, 2006.

6.2 Electrolytic Processes

6.2.1 Outline of Process Techniques

Electrolytic processes are a function of the type of feed used. Two cases are given:

First—Extraction from dolomite and magnesite ore: Dolomite is converted to magnesium chloride by using the following steps:

- Crushing and roasting of dolomite
- Mixing with seawater in large tanks, where magnesium hydroxide settles to the bottom
- Heating, mixing in coke, and reacting with chlorine, producing molten magnesium chloride that can be electrolyzed to produce magnesium, which floats to the surface

Second—Extraction from sea salt: Sea salt and salt brines are used as a raw materials to produce magnesium by electrolysis. Normally, these sources contain about 10% magnesium chloride, a highly significant content together with an appreciable amount of water. Before electrolysis, magnesium chloride has to be dried to make it anhydrous magnesium chloride. This is a primary condition before it can be electrolyzed to produce metal.

The plant at Dow Chemicals in Freeport, Texas, represented the first facility to produce magnesium metal extracted from seawater. The Freeport facility operated until 1998.

At present, the Dead Sea Magnesium Ltd. is the only facility worldwide that produces magnesium from seawater.

6.2.2 Fundamentals

Fused salt electrolysis is an attractive technology for the manufacture of magnesium. If the source of electricity is a renewable energy resource (wind and solar), this route could be further pursued. Molten salt electrolysis of magnesium chloride represents a primary role in this industry.

Talking about sustainable development along this line of fused salt electrolysis, the challenges to be faced lie within two key elements:

1. Improvement of existing technologies
2. Development of novel technologies, both having much less environmental impact

For metals extraction and materials processing, the latter element can potentially lead to the displacement of many conventional non-electrochemical technologies used today. As a matter of fact, molten salt electrolysis technology as applied for the earlier applications remains undeveloped for the last decade. Significant development and changes are cited in recent years, all the way from lab scale to commercial scale through pilot plant experimentation.

The following is a *fundamental explanation* for the basic reactions that take place in the electrolysis of molten salts. A typical example is the electrolysis of sodium chloride to produce caustic soda. Ionic binary compounds, by definition, contain only two elements, a metal and a non-metal. When the ionic compound is in the molten state, the locked ions within the ionic structure will be free to move about (conducting electricity).

A diagram for the electrolysis of molten compounds is shown in Figure 6.1.

To conduct electricity within the structure of sodium chloride, enough heat has to be supplied to form a molten compound. When electrolysis takes place, the separation of the molten compound into elements occurs. The reaction products of the salt at each electrode are called, by definition, half equations.

The chemical equations are written for each of the half equations as follows:

$$2Na^+ + 2e^- \rightarrow 2Na, \text{ releasing sodium metal at the } (-) \text{ cathode}$$

$$2Cl^- - 2e^- \rightarrow Cl_2, \text{ releasing chlorine gas at the } (+) \text{ anode}$$

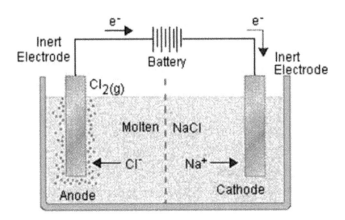

FIGURE 6.1
Illustration of electrolysis of molten salts.

It is observed that a balance in electrons occur, where the same number of electrons show in each equation. This is explained by the following: Sodium atoms are produced from the ions by gaining two electrons (reduction process). On the other hand, chloride ions lose electrons, forming chlorine atoms (oxidation).

A molecule of chlorine gas is now formed by the combination of chlorine ions. The overall reaction will be given by the equation:

$$2Na^+Cl^-_{(l)} \rightarrow 2Na_{(s)} + Cl_{2(g)}$$

6.2.3 Electrolysis of Magnesium Chloride

The two principal methods for the manufacture of anhydrous magnesium chloride are:

1. Dehydration of magnesium chloride brines
2. Chlorination of magnesium oxide

The method is summarized as follows: The feed to the electrolytic cell is made of granules of anhydrous magnesium chloride. Normally, it is important to have a fused salt mixture of the following ingredients: $MgCl_2$, $CaCl_2$, and $NaCl$ in the ratio of 20%, 20%, and 60%, respectively, to be maintained in the cell. A high-amperage direct current is used to supply the necessary power. For the current passage, cells are made of cathodes and anodes. The cells themselves, made of steel, represent the cathodes, while the anodes, on the other hand, are made of artificial graphite.

From the power point of view, conditions of electrolysis require 6–7 volts and 100,000 amps. Once electrolysis is finished, the produced molten magnesium finds its way to the top of the cell. The next step is then to collect it and process it for casing into ingots. The purity of the magnesium product is about 99.8%. Magnesium metal is put into market by making an alloy with zinc or other metals.

Per one ton of magnesium to be produced, the raw materials and the energy (both electricity and fuel) requirements are as follows:

First: Raw materials (magnesium chloride): 4.2 tons

Second: Fuel (natural gas): 36,000 cu.ft

Third: Energy (electrical power): 18,500 Kwh

Fourth: Others (materials for electrodes): 0.10 tons

Other options for energy supply for the electrolysis of magnesium chloride are the following:

- Solar power and wind energy (renewable energy sources)
- Off-peak electricity

The following are some of the common types of electrolytic processes in use:

- Dow Chemical Process
- Norsk Hydro Process
- National Lead Industries Process

Chlorination of magnesite, forming magnesium chloride, is carried out, on the other hand, by:

- IG Farben Process
- MagCan Process

6.2.4 Types of Electrolytic Cells

The electrolytic cell is a vital component for the magnesium industry. Different types are available in the market. The Dow cell and the IG Farben cell are the two most well-established cells industrially. The most distinguishing feature of the Dow cell is that it has an externally heated rectangular steel pot; on the other hand, the IG Farben cell is lined with insulating refractory brick contained in a steel tank. No provision exists for external heating.

Other cell types that are commercially known are:

- Alcan cell, developed by Alcan from Canada
- VAMI cell, developed by VAMI from Russia. It utilizes the same principles as that of Norsk Hydro
- Ishizuka cell, developed by the Ishizuka Research Institute from Japan

6.2.5 Dow Process

Details of the Dow process are given as a model for a commercial electrolytic process. This is best described as shown in the flow diagram in Figure 6.2.

In brief, the feed into the cell is made of anhydrous magnesium chloride (Figure 6.3). The cell has to be kept hot enough to allow the feed to melt. Once electrolysis has been completed, magnesium and chlorine are produced from the cell as end products, as shown next:

$$+\text{Anode: } 2Cl^- \rightarrow Cl_2 + 2e$$

$$-\text{Cathode: } Mg^{2+} + 2e \rightarrow M$$

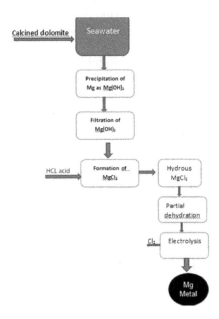

FIGURE 6.2
Dow process for the production of magnesium by the electrolytic method.

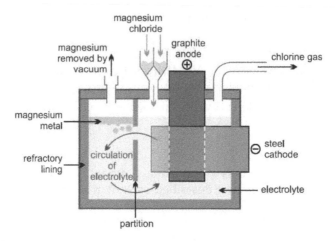

FIGURE 6.3
Electrolytic cell for the electrolysis of magnesium chloride.

6.2.6 Production of Magnesium Chloride from Magnesium Oxide

The naturally occurring carbonate of magnesium, known as magnesite ($MgCO_3$), represents an important natural source for the production of magnesium chloride, through the conversion first to magnesia (MgO).
The process is described by the following steps:

1. By heating the dolomite to a high temperature, conversion to mixed oxides takes place.
2. Seawater is treated first with the dolomite, obtained from the previous step.
3. Magnesium hydroxide is recovered as precipitate by filtration; when heated, pure magnesium oxide is readily formed.
4. Finally, magnesium chloride is obtained from the oxide by reducing the hot oxide using carbon in a stream of chlorine. Chlorination takes place by maintaining a high temperature using an electric furnace (Figure 6.4).

The main reactions leading to the formation of magnesium chloride are as given by the next family of equations:

$$\text{First: } 2MgO + C + 2Cl_2 \rightarrow 2MgCl_2 + CO_2$$

$$\text{Second: } Cl_2 + C + H_2O \rightarrow 2HCl + CO$$

$$\text{Third: } MgO + 2HCl \rightarrow MgCl_2 + H_2O$$

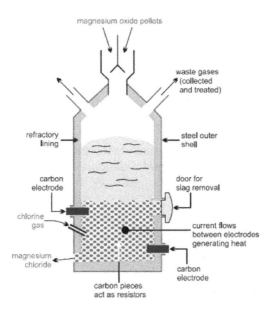

FIGURE 6.4
Formation of magnesium chloride from magnesium oxide.

If magnesium chloride-rich brines are to be used as a feed to produce magnesium, some preparation steps are to be carried out. Impurities in the solution have to be removed first. Next, a concentration process by evaporation in stages is done. In order to avoid hydrolysis of the magnesium chloride while concentration is carried out, the last step is done in the presence of hydrogen chloride, as shown by the following chemical equation:

$$Mg(OH)Cl(s) + HCl(g) \rightarrow MgCl_2(s) + H_2O(g)$$

6.3 Thermal Reduction Process

As the name implies, thermal processes require the usage of high heat. Over the past period, many well-known processes in this field have been in use for magnesium production.

The process necessitates the usage of a reducing agent. Silicon is the main choice while reduction of magnesium oxides takes place under a vacuum. The main sources of magnesium for this process are: dolomite ($MgCO_3 \cdot CaCO_3$) and magnesite ($MgCO_3$). Consequently, these two ores represent the source of the magnesium oxides to be used in the thermal process, as will be explained next.

On the top of the list of thermal processes comes the one developed by Pidgeon (1943). The process found applications in more than one place. This includes the United States, Canada, India, Japan, and the United Kingdom. A very large magnesium production capability based on the Pidgeon process was developed by China.

About 80% of the total production worldwide is attributed to China using the silicothermic Pidgeon process. That makes China to be recognized as the dominant supplier of magnesium.

The question is posed: Why is the Pidgeon process known as an energy- and labor-intensive form of thermal reduction? The answer is because the process is described by the following features:

- Dolomite is the main ore.
- Ferrosilicon is the reducing agent.
- Coal is the energy source.

These unique features mark this process as the main production method for magnesium in China.

Process description:

In this process, a mixture of calcined dolomite ore and ferrosilicon is introduced to closed-end, nickel-chromium-steel alloy retorts. Heating of these filled retorts is continued until magnesium crowns are formed. The operating time per cycle is about 11 h. Manual filling and emptying has to be done. The raw material used is estimated to be about 11 tons per 1 ton of magnesium produced.

The widespread use of the Pidgeon process, particularly in north-central China, is attributed to the use of the coal in these provinces, which are located in an area known as a *coal-rich* region. This region also enjoys lower costs for labor and energy. By far, it is much lower compared to other magnesium producing regions.

Because of the relatively cheap electricity, electrolysis was predominantly used as the method of production in the country. Then, it was displaced by the thermal technology. However, the use of an old version of the thermal reduction process in most of the Chinese plants is common.

The steps involved in the *thermal* process in general are described as follows:

First: Calcination of dolomite is carried out for the crushed ore by heating in a kiln, producing a mixture of magnesium and calcium oxides. This reaction is given by the following equation:

$$MgCO_3 \cdot CaCO_3(s) \rightarrow MgO \cdot CaO(s) + 2CO_2(g)$$

Second: This step is concerned with the reduction of the magnesium oxide to magnesium.

Third: Preparation of the reducing agent is done by heating sand and scrap iron to produce an alloy known as ferrosilicon that contains about 80% silicon.

Fourth: Crushed ferrosilicon, obtained from third step, is mixed with magnesium oxides. This combined mixture is shaped into briquettes. Then, it has to be loaded into the reactor. In order to reduce the melting point of the slag, alumina (Al_2O_3) may be added to the mixture described before.

Fifth: The operating conditions for the reaction are: A temperature in the range of 1500–1800 K, and a very low pressure (close to vacuum). Magnesium is produced as a vapor. This vapor has to be changed into a liquid. Cooling to about 1100 K in steel-lined condensers will bring the hot vapors to condensation, where it is removed and cast into ingots.

$$2MgO(s) + Si(s) \rightleftharpoons SiO_2(s) + 2Mg(g)$$

The total reaction as given by the earlier equation is represented by the forward reaction and the opposite one. The forward reaction, producing magnesium, is endothermic and is favored by heat input. The reverse reaction, on the other hand, goes for magnesium oxide. However, the net equilibrium position will be in favor of magnesium production. Another added feature in favor of magnesium production is the removal of the produced magnesium. This will make the reaction shift to the right and goes to completion.

The produced silica forms calcium silicate, or molten slag, as it combines with calcium oxide:

$$CaO(s) + SiO_2(s) \rightarrow CaSiO_3(l)$$

High purity product (about 99%) is obtained from the ferrosilcon. When compared with the electrolytic process, the ferrosilcon is superior by small margin.

When the reduction process is compared to the electrolytic approach, one finds high purity metal (99.95% Mg) is produced by the former method. This is true if the thermal process is carried out that under the right conditions. Dolomite, and to a lesser extent, magnesite, represent the main ore minerals in the production of Mg metal by thermal reduction. This is illustrated as presented in Table 6.2.

TABLE 6.2

Comparison of Mg Metal Production Methods

Method	Electrolysis	Thermal Reduction
Raw materials	Magnesite, dolomite, bischofite, carnallite, serpentine, olivine, seawater, and brines	Dolomite, magnesite
Energy type	Hydroelectric, gas, and oil	Coal, gas
Energy requirements (MWh/tons of Mg)	18–28	45–80
Process	Continuous	Batch
Capital investment (US $/ton of Mg capacity)	10,000–18,000	≤2000
Manpower	X	5X

Source: Magnesium—Raw Materials, Metal Extraction and Economics—Global, http://www.em.gov.bc.ca/Mining/Geoscience/IndustrialMinerals/Documents/Magnesium.pdf (accessed June, 2016).

6.4 Magnesium Processes for Tomorrow

The search for more sustainable processes for magnesium production is underway. Using the case of the Pidgeon process as an example, it is clearly found that low productivity, high labor requirements, and high energy consumption are some of the typical drawbacks. Moreover, the ferrosilicon reactant used in the process as a reducing agent involves huge energy consumption. This marks the Pidgeon process as an unfavorable approach in terms of high Global Warming Potential (GWP).

Therefore, the motivation for the development of more sustainable methods for magnesium production is highly commended and much attention has been received worldwide.

The following are some typical examples for much more refined and newly introduced methods for the production of magnesium. They are classified into the well-known two routes:

6.4.1 First: Electrolytic Processes

This group includes the following:

6.4.1.1 Process 1—Pacific Northwest National Laboratory

Description: It is a proposed new process to produce magnesium from seawater. The current methods applied to manufacture magnesium from seawater as a raw material suffer from being energy intensive and expensive. The main reason is the fact that the concentration of magnesium chloride is rather low in seawater. Energy is spent in the extraction of magnesium as end product, either by precipitating magnesium as hydroxide and then converting it to magnesium chloride, or by obtaining the magnesium chloride directly by preferential salt separation, known as PSS. Further, the chemical bond between magnesium and chlorine consumes energy. This is because it requires preheating the feed (salt) to a temperature about 900°C–1000°C and then applying the electric current to break the bond.

Pacific Northwest National Laboratory's (PNNL's) new process introduces much milder conditions. It operates at a low temperature. In addition, a low-energy dehydration technique is used in the process. That step is combined with a new catalyst-assisted process. The basic function of such a catalyst is to formulate an organometallic compound. This is simply a combination of the catalyst with a reactant, that is, magnesium chloride. To obtain magnesium as an end product from the formed organometallic compound, it is broken at temperatures less than 300°C by using a proprietary process. This procedure replaces the electrolysis step for magnesium chloride salt, eliminating it completely. The advantages brought in by this proposed process are immense. This may result in appreciable savings in overall economics. In addition, a more efficient operation than the conventional magnesium extraction method is obtained.

Main features: From what was presented, many advantages are credited to PNNL process. It is based on introducing a low-cost, low-energy, modified, new metal-organic compound. This labels a new process for producing magnesium from seawater. This process, when compared with the conventional one, eliminates many of the energy intensive steps.

In addition to the savings obtained in terms of money, other tangible benefits are achieved as well from the environmental point of view. It is anticipated to obtain about a 50% reduction in the energy consumed. This will result in a substantial decrease in CO_2 emissions over conventional methods.

Seawater is considered, by far, an abundant source of magnesium. With this proposed technology, which is based on an economic and environmentally sound method, a worldwide supply of magnesium can be made available for centuries.

6.4.1.2 Process 2—Solid Oxide Membrane Technology (MgO)

Description: A rather a new approach has been pursued to obtain magnesium from MgO, which involves the electrolytic reduction of MgO, known as Solid Oxide Membrane Technology.

Reduction takes place, where, at the cathode, gaining two electrons by the Mg^{2+} ion causes its reduction to magnesium metal.
The anode is a liquid metal and the electrolyte is Yttria-stabilized
zirconia, known as YSZ. With a layer of graphite bordering the liquid metal anode, oxidation takes place at the YSZ/liquid metal
anode. Carbon and oxygen react to form carbon monoxide at
this interface.

   If silver is used as anode, gases evolved at the anode contain
oxygen only.

Main features: When compared with the electrolytic reduction process
from the economic point of view, a substantial reduction of about
40% in cost per pound could be obtained. Again, it is a friendlier
method from the environmental aspect. Compared to others, the
carbon dioxide emitted from the new process is less.

Chemical reactions:

$$2MgO(s) + C(s) + 2Cl(g) \rightarrow MgCl_2(s) + CO_2(g)$$

$$Cl_2(g) + C(s) + H_2O(g) \rightarrow 2HCl(g) + CO(g)$$

$$MgO(s) + 2HCl(g) \rightarrow MgCl_2(s) + H_2O(g)$$

6.4.1.3 Process 3—Norsk Hydro Process

Description: In the field of the magnesium industry, one would count
Norsk Hydro as one of the major magnesium producers. Their
contribution in developing the magnesium production process
is well-known worldwide. When one considers a comprehensive
research review on the electrolytic processes of magnesium production, it reveals a striking fact that changes and developments
accomplished in this area are not much, and have been this way
for some time.

   Two well-known processes are credited to Norsk Hydro, as far as
the type of feed:

   1. One in which magnesite is the feedstock
   2. The other in which seawater and dolomite are the feedstock

Main features: Advancement and developments took place both in the
feed preparation technology and the cell sizes.

   Attaining a higher income in terms of return on its capital investment could create an incentive for Norsk Hydro to offer their latest
technology for commercial licensing.

6.4.1.4 Process 4—Dow Modified Cell

Description: It utilizes two new additions: using a feed that contains water of hydration (a hydrous feed) and installing larger anodes. These features will increase production and enhance the efficiency of the Dow cells, modified from their original type. In addition, they managed to cut the energy spent for magnesium production by 30% by adopting a very well-balanced procedure. The observation was then made when the plant was shut down that the cells were performing with a current efficiency of close to 90%, at a capacity of 200,000 amps. This represented a significant rise from the initial loads on the cell. A decision was made by Dow to quit the magnesium business. A license was granted to Pima's 100%-owned subsidiary for the improved magnesium production technology.

Main features: Changes in the design of the process were implemented to replace the feedstock to the cells; rather than using seawater, high-grade magnesite is to be used. This is different from the old process Dow had used in Freeport, Texas.

6.4.1.5 Process 5—Dead Sea Magnesium

The establishment of a new technology was accomplished at the Dead Sea for the design, construction, and the establishment of the magnesium production plant using Russian/Ukrainian technology. In this project, new technology was introduced as far as feed preparation and other aspects. It included the latest of Commonwealth of Independent States as well.

6.4.2 Second: Thermal Processes

Description: The thermal approach represents the ideal method for magnesium production. The direct reduction of calcined magnesite with carbon attracted the attention of many researchers through experimentation using pilot plants, which were developed to a commercial scale. However, results were not satisfactory and it was not well-received by the industry.

In one of the experiments, a mixture with a composition 76% of magnesia and 24 parts of high-petroleum coke (highly volatile type) was used. Further, this mixture was introduced to be ball-milled and pelleted. The process was further extended to charge the mixture in a reduction furnace (3-phase type furnace) at an operating temperature of 1950°C–2050°C.

Main features: The products obtained are magnesium vapor and carbon monoxide, which requires a special technique to separate them from each other. The basic idea behind separating the two products is based on fast cooling to avoid recombination. This was accomplished by proposing a stream of cold hydrogen, causing a kind of a *shock* cooling. As a result of this type of cooling, magnesium metal condenses out in the form of a fine powder (pyrophoric).

6.5 Magnesium Recycling

Recycling means converting waste into reusable material, or reuse and reprocess. Although such a topic is outside the domain of this book, a short note is presented on magnesium recycling. The fabrication of magnesium alloy parts is basically done through die casting. It is the most common method used. Three different options are available to manage the amount of process scraps by the die-casting foundries.

The following are some important parameters to consider in die-casting:

1. The ratio of scrap to product for a single shot is the number one parameter.
2. The amount of material lost in the melting cycle has to be taken into account.
3. The quantity of different components that are part of the cast, as well as the percentage of cast parts that must be rejected during production.
4. Other important parameters are: the end quality of process scrap, and the recycling operation efficiency. All of these factors affect the amount of process scrap and primary magnesium utilized.

Some types of magnesium alloys can be recycled back into products, including magnesium used in structural applications. The recycling process could display the same properties the primary magnesium metal had. This includes chemical, physical, and mechanical characteristics. When it comes to the energy spent in recycling magnesium alloys, there is a substantial reduction in the energy consumed compared with that used in primary production. It is only 5% of the energy required to produce primary magnesium alloys.

6.6 Annual Production of Magnesium

According to the references cited in the table, the following data are given. These figures are for *primary* production from the ore and do not include secondary production from recycled materials.

World	910,000 tons[1]
China	800,000 tons[1]
U.S.	70,000 tons[2]
Russia	30,000 tons[1]
Israel	25,000 tons[1]
Kazakhstan	20,000 tons[1]

Data from:
[1] U.S. Geological Survey, Mineral Commodity Summaries, 2016.
[2] Last available figure is 2012 (Minor Metals Trade Association (https://mmta.co.uk/).

The quantity of recycled magnesium is estimated to be 3% of the total magnesium used annually. It is estimated to be about 23,000 tons.

Finally, a comparison between the well-established production methods for magnesium and recent ones is shown in Figure 6.5 based on the production costs in $/lb. The recent ones include the *carbothermic* method and *SOM* method, which are briefly described next:

The *carbothermic* reduction of magnesium oxide using a carbon as a reducing agent has been introduced as a recent technique. The method has drawn attention because it is much cheaper than ferrosilicon. Moreover, it is anticipated to have lower magnesium production costs because of many factors, which include:

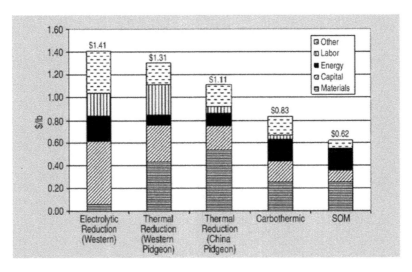

FIGURE 6.5
Production cost of major primary magnesium production methods. (From JOM, The Member Journal of The Minerals, Metals and Materials Society, https://www.tms.org/JOM.)

1. Reduced equipment size; consequently, lower capital costs
2. Cheaper reductant used
3. More efficient productivity than that achieved through electrolytic routes

However, the process has not yet been commercially developed because of the back-reaction between the magnesium vapor and carbon monoxide.

The solid oxygen-ion conducting membrane (SOM) process for magnesium production, on the other hand, considered by Safe Hydrogen, is based on passing a current through a flux containing magnesium oxide. This flux acts as the cathode since it is contained in a stainless-steel crucible. The oxide reduction in this process involves a simple magnesium hydroxide-calcining operation. It is electrochemical in nature but has the advantage of replacing the intensive magnesium chloride dehydration process, which is a prerequisite for the conventional magnesium electrolytic process.

7

Current Applications of Magnesium and Prospective Ones in the Field of Energy Domain

7.1 Introduction

From the structural point of view, magnesium is considered the third most commonly used structural metal, following iron and aluminum. When someone searches for magnesium as a source of energy, he finds many publications on its use as a biological agent, providing energy for the human body. The following statements are some typical examples:

- Magnesium can help treat the flu.
- Magnesium can boost your energy and improve sleep.
- Magnesium in the body is your true energy source.

In brief, magnesium is an essential mineral nutrient (i.e., element) for life and is present in every cell type in every organism. In addition, magnesium is a constituent of the chlorophyll in green plants and is necessary in the diet of animals and humans.

We need to consider magnesium as a source of power, knowing that *energy* can be stored, whereas *power* cannot be stored. *Energy* comes with a time component; *power* is an instantaneous quantity.

Can this dream be realized? Fortunately, a number of publications and books are on the spot that capture the reader's interest in using magnesium as a source of power for our global needs. Examples of some titles that appeared in these publications are cited next:

> *"New Power Sources Could Be Made Using Magnesium"*
> *"The Magnesium Civilization: An Alternative New Source of Energy to Oil"*
> *"The Magnesium Miracle: Could Magnesium Save the World?"*
> *"Clean Magnesium Energy Cycle Hints at Fossil Fuel Freedom"*

7.2 Current Applications of Magnesium

In the field of its applications as a structural material, magnesium is a commercially important metal with many uses. It is only two-thirds as dense as aluminum. Magnesium applications, in general, are motivated by its light weight, high strength, and high damping capacity. In addition, close dimensional tolerance and ease of fabrication of its alloys are advantageous. An appreciable increase in these applications and great strides in development in this field took place in the twentieth century. The metal and its alloys are increasingly in demand, promoting the magnesium industry along this line.

7.3 Magnesium Alloys

Basic definition, cited by Wikipedia: "An alloy is a mixture of metals or a mixture of a metal and another element. Alloys are defined by a metallic bonding character especially to give greater strength or resistance to corrosion. An alloy may be a solid solution of metal elements or a mixture of metallic phase."

When it comes to magnesium alloys, they are mixtures of magnesium with other metals, usually aluminum, zinc, manganese, silicon, copper, rare earths, and zirconium. Practically, all the commercial magnesium alloys manufactured in the U.S. contain 0.1%–0.4% magnesium and 3%–13% aluminum.

Magnox alloy, on the other hand, whose name is an abbreviation for *magnesium non-oxidizing*, is 99% magnesium and 1% aluminum. It is used mainly in the cladding of fuel rods in magnox nuclear power reactors. Magnesium alloys are referred to by short codes defined in ASTM B275. This acronym denotes approximate chemical compositions by weight. As an example, an alloy with the code "AS41" has 4% aluminum and 1% silicon. If aluminum is present, a manganese component is almost always also present at about 0.2% by weight, which serves to improve grain structure; if aluminum and manganese are absent; zirconium is usually present at about 0.8% for this same purpose.

Magnesium alloys are used in wrought form including extruded bars, sections, and tubes and also forgings and rolled sheets. Several wrought alloys are based on the Mg–Al–Zn system; if aluminum and manganese are absent, zirconium is usually present at about 0.8% for this same purpose.

Considering magnesium alloys as a replacement for some engineering plastics, they find a variety of applications. This is basically due to their higher stiffness, high recycling capabilities, and lower cost of production.

7.4 Mechanical Applications for Magnesium

Production of *light* cars using magnesium was the focus of interest by the industry for many years. Magnesium found many applications in different areas. Beside its original use for die-cast parts, other methods of fabrication are pursued in other areas; in particular, rolling and stamping, as well as extrusions. This is a good indication that the future of magnesium in automotive usage is expected to continue to grow at a rapid pace.

The main targets for automakers worldwide could be simply stated as follows:

- Better fuel economy
- Reduced emissions

These factors boost the use of magnesium in die-castings. It is expected to grow at an appreciable rate of 10%–15% per year.

When it comes to weight reduction without compromising overall strength, then magnesium alloys are the preferred material. Other mechanical properties such as vibration damping capacity are considered in many applications. This is very essential where the internal forces of high-speed components must be reduced.

Illustrations for the application and use of magnesium are given in Figure 7.1.

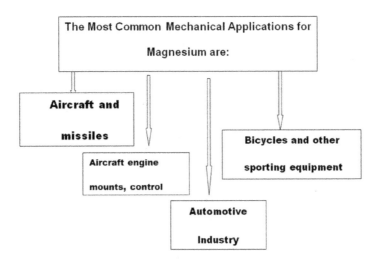

FIGURE 7.1
Mechanical applications for magnesium.

Magnesium alloys for automotive applications require a combination of high strength properties and low density. In addition, weight reduction will improve the performance of a vehicle by reducing the rolling resistance and energy. This will reduce the fuel consumption, hence, a reduction in the greenhouse gas CO_2. This improves the environment.

Other appreciable benefits are accomplished by the use of magnesium in automotive applications beside weight savings. There has been a strong motivation for many years to use magnesium in the front end of a vehicle. This will provide not just a lower overall mass for the car, but will cause the shifting of the center of gravity toward the rear. This will improve the car's handling and turning capabilities.

Magnesium can also replace steel components in vehicles by using a single cast piece of magnesium. This adds to the strength of the material and allows for housings to be cast into place.

According to the United States Automotive Materials Partnership (USAMP), it is estimated that by 2020, and *per one vehicle*, 350 lbs of magnesium will replace 500 lbs of steel and 130 lbs of aluminum. Weight reduction of 15% is anticipated, which in turn will lead to fuel savings of 9%–12%.

Based on the fact that a large number of vehicles are produced worldwide, a significant drop in the carbon dioxide released into the atmosphere will be in effect because of the weight savings in the vehicles. The elimination of carbon dioxide will reduce its harmful impact on global warming. This entails magnesium to be called the "green metal of the twenty-first century."

At present, the automobile industry produces many parts made from magnesium alloy for their needs. Examples are: engine block, wheels, steering columns, seats, front consoles, and hoods.

Magnesium usage is extended further to a new and very promising field that is the *sports* sector. It utilizes an extra unique benefit to magnesium; besides being lightweight, durable, and stronger than aluminum, it has an absorbing property of 16 times more for shock and vibrations.

As for *Aerospace Applications*, magnesium has been around in the aerospace industry for long time. The metal has been used in many applications, both civil and military. To reduce the weight of air and spacecraft, as well as projectiles, is highly significant in this kind of industry. This will lead to a decrease in emissions and boost fuel efficiency and will reduce operational costs as well.

To cite some specific applications in the *Aerospace Applications*, magnesium is used in the thrust reversers for all family members of Boeing as well as in engines and aircraft transmission casings.

Spacecraft and missiles applications: Lift-off weight reduction is a very crucial factor to consider in the design. Magnesium alloys would serve the purpose. It provides a material that can withstand the extreme conditions faced during the craft or missile operation. Extreme elevated temperatures, exposure to ozone, and the impact of high energy particles and matter are additional properties that magnesium alloys add to a system.

Magnesium, when infused with silicon carbide nanoparticles, it has extremely high specific strength. This property finds applications to be used in super-strong, lightweight materials and alloys.

Finally, and most importantly, it is used in preventing the corrosion of iron and steel, as in pipelines and ship bottoms (sacrificial applications). For this purpose, a magnesium plate is connected electrically to the iron. The rapid oxidation of the magnesium prevents the slower oxidation and corrosion of the iron. This is known as sacrificial cathode protection.

7.5 Other Applications

For *Medical Applications*, there is a long list to be presented. Initially, magnesium was first introduced in the medical field as an orthopedic biomaterial.

Another application is the use of magnesium to store hydrogen as magnesium hydrides. Chapter 8 is devoted to this topic.

Now, the most notable increases in consumption in magnesium have been mainly in *non-energy applications*. It is worthwhile to mention that the restricted use of magnesium in energy applications is basically due to unfavorable economics, not because technical limitations.

It is proposed to say that the economics of magnesium could be very well enhanced by considering the following two factors:

- Development in the extraction technology of magnesium
- Finding new avenues for technical application of magnesium to provide power in the energy domain

According to the statement made by Takashi Yabe of the Tokyo, "there is enough *magnesium* to meet the world's *energy* needs for the next 300,000 years." This could be a very good incentive to pursue this line of using magnesium in energy applications.

Next, a number of projects related to *magnesium-energy applications* are presented. A more elaborate presentation on this topic is beyond our scope.

7.6 Magnesium as Energy Carrier

7.6.1 Background

The search for appropriate means to store energy is one of the biggest obstacles to the widespread adoption of alternative renewable energy sources. There is more than one option. Using batteries is well practiced, but they can be bulky. Many proposals have been made for energy systems that would utilize solar energy to split water, producing hydrogen. This establishes the fact that solar energy has been stored in the form of hydrogen. Hydrogen, in return, represents a reliable, ideal energy carrier.

Now having hydrogen available as energy carrier, the question then arises: how do we store and transport hydrogen to end users?

The answer could be the use of magnesium, which would save the planet by capturing sources of alternative energy to replace fossil fuels. This is detailed next in a number of projects.

7.6.2 Process Description for Magnesium as Energy Carrier

Renewable energy, such as solar energy, is used as the energy source to make pure magnesium from sources like seawater. Our proposed energy scheme involves using solar to vaporize a dynamic stream of pre-concentrated seawater (brine) flowing along an inclined Preferential Salt Separator, PSS (Abdel-Aal & Coworkers). This evaporation process takes place by direct solar radiation or by using a heating system powered by a photovoltaic source. Magnesium chloride salts, soluble in seawater, will separate as end product. Fresh water will be produced as a by-product as well. Anhydrous magnesium chloride is collected and then electrolyzed to produce magnesium metal using energy generated by solar power. Once produced, magnesium represents a reliable source of stored energy that could be exported by air, sea, or other means of transportation to remote locations for power-generation. Magnesium can be used on-site to construct a galvanic cell that consists of magnesium/iron electrodes generating electricity. Therefore, when connected to a less reactive metal, the magnesium becomes the anode of an electrical cell and corrodes in preference to the other metal. The produced electric current is fed to the electrolytic cell to electrolyze water to produce hydrogen. Another option is to use magnesium as a storage medium to store hydrogen in the form of magnesium hydride (this is discussed in detail in Chapter 8).

7.6.3 Design Features of the Proposed Energy System

The system utilizes solar energy to its optimum operating capacity all components, which include the following units:

- Desalination and hydrogen production by water electrolysis
- Preferential Salt Separation (PSS)
- Electrolytic cells for magnesium production
- Mg/Fe galvanic cells for hydrogen generation

These components are illustrated as given in Figure 7.2. The PSS operation was described earlier in Chapter 5. The diagram given in Figure 7.3 describes the overall process.

Hydrogen production using galvanic-electrolytic cell is illustrated in Figure 7.4. The dissociation of water is a non-spontaneous reaction and needs energy to split water-producing hydrogen and oxygen (4.8 kWh/one cubic meter hydrogen). This energy is supplied in situ by another spontaneous reaction using galvanic cells.

7.6.4 Process Description for Recycling Magnesium

Recyclying magnesium involves a renewable energy cycle based on magnesium and a solar-energy-pumped laser, after (T. Yabe). The solar energy is

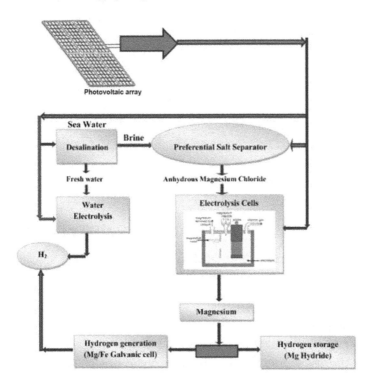

FIGURE 7.2
Overall proposed energy scheme for magnesium as energy carrier.

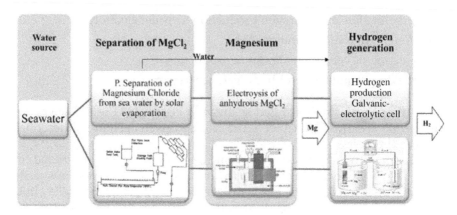

FIGURE 7.3
Seawater as a source of magnesium metal.

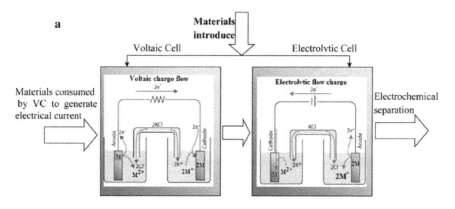

FIGURE 7.4
Functions of galvanic cells and electrolytic cells to produce H_2.

used to recycle the used magnesium, (MgO) to recover new magnesium, and magnesium is used as fuel-cell battery.

As mentioned before, several candidates exist for storing the solar energy. It is important to take into consideration the following conditions for a candidate for storing solar energy:

- It must be transportable.
- The employed materials must be abundant.
- The process should be friendly to the environment.

The problem is how to transfer energy from one place to a remote area where energy is needed. Using the electrical cable for a long-distance power supply

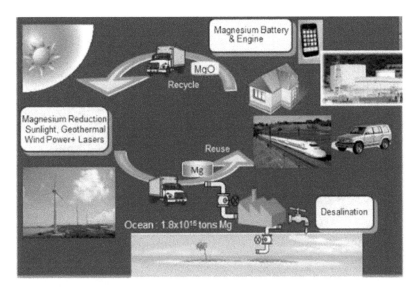

FIGURE 7.5
An overall scheme for recycling magnesium.

crossing several countries is not practical because of huge construction cost and national security. Hydrogen might be one possibility; however, energy density per volume is very small compared with gasoline, even in the liquid form.

Therefore, it has been proposed to use magnesium as a storing material because it is abundant. The reaction product is MgO or $Mg(OH)_2$, which are both harmless and in a solid state that can be easily collected. The only problem was knowing how to produce and recycle it. It was then proposed to use lasers to recycle magnesium oxide as demonstrated in Figure 7.5.

7.7 Magnesium as an Alternative Power Source

There is a great potential for the application of magnesium in fuel cells. For the sake of comparison, cars using Zinc-air-fuel cells achieved 600 km mileage in 2003. When using Magnesium-air-fuel cells, 3 times more energy is produced, which is 7.5 times more effective than Lithium-ion batteries.

It has been reported by Dr. Yabe that a high-temperature solution was achieved by concentrating solar collectors and a solar-pump laser. A temperature as high as 3,700°C was obtained. As far as its application in magnesium industry, this high-heat method is used to burn magnesium oxide extracted

FIGURE 7.6
Pure magnesium can be used as a fuel.

from seawater. The solar-pumped laser is necessary to help obtain this high temperature.

The most important work on this was published by Professor Yabe in his book entitled *The Magnesium Civilization: An Alternative New Source of Energy to Oil*. If applied, this could be the answer to the world's energy needs.

Other interesting research topics are the following:

- "Magnesium and the Magnesium Power Engine." Pure magnesium (Figure 7.6) is used as a fuel. When magnesium is burned as an element, its energy density is so great, it is about ten times that of hydrogen.

- Another interesting application to utilize magnesium to provide energy is to mix magnesium with water. This reaction produces heat, boiling the water and converting it to steam. The produced steam can then be used to drive a turbine and do useful work. The reaction also produces hydrogen as a result of water dissociation. Hydrogen adds an additional advantage, as it can be burned to produce even more energy. The by-products are simply water and magnesium oxide. The latter can then be converted back into magnesium using the solar laser. The trouble is that concentrated solar collectors tend to be huge and costly, and solar-pumped lasers are normally very low-powered. The solution is using relatively small Fresnel lenses, transparent and relatively thin planar lenses made up of concentric rings of prisms, which are commonly found in lighthouses to magnify light in a way that would normally require a much larger, thicker lens. It normally only absorbs about 7% of the energy from sunlight, but when doped with chromium, this figure increases to more than 67%.

- The chemical reaction between magnesium (in a powder form) and water at room temperature produces high-energy steam and hydrogen. The hydrogen is burned at the same time to produce additional high-energy steam. The energy cycle produces no carbon dioxide or other harmful emissions. An experiment was performed to prove that the revolutionary Magnesium Energy Cycle could one day free society from its dependency on fossil fuels.

- A metal-air cell was devised by engineers at *MagPower* that uses salt water and ambient air to react with a magnesium anode generating electricity (Figure 7.7).

- Another additional application for magnesium is to use it in a based version of the lithium-ion rechargeable cell. It is a type of battery known for its long life and stability. It would be ideal for storing electricity from renewable sources.

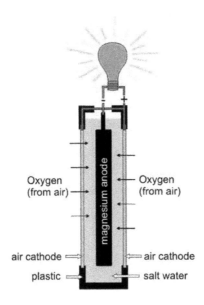

FIGURE 7.7
Magnesium + oxygen + water + salt + additive = direct current. (From MagPower Systems, Inc., www.magpowersystems.com/.)

- On-board hydrogen generation for cars by reacting magnesium with steam was carried out by Andrew Kindler[*] at the California Institute of Technology at Pasadena. Pure hydrogen is used in fuel cells to generate electricity, leaving behind only magnesium oxide, a relatively benign material, as a by-product.

- Low-cost magnesium energy cells, *metal-air cells*, have been developed by *Aqua Power*. This represents a potential for high energy generation without hazards. It represents a low-cost magnesium energy cell with a power output doubling that of lithium ion batteries. Realistic Magnesium Air Fuel System (RMAF) fuel cells have a very long shelf, are lightweight, transportable, environmentally friendly, safe, and easily scalable for greater power generation. Aqua Power has aggressively patented (16 patents and patents-pending to date) and actively protects its intellectual property in Japan and internationally.

7.8 Magnesium and Global Warming

Metallic magnesium is highly reactive and stores a lot of energy. Even a small amount of magnesium ribbon burns in a flame with a satisfying white heat (see Appendix 1). It burns white hot underwater as well as in a pure carbon dioxide atmosphere. Now, when it comes to CO_2 atmosphere, the burning magnesium with its terrific heat splits the CO_2 molecule apart into C and O_2 (*Bruce Mulliken*)[†]. With oxygen available, it combines with magnesium forming magnesium oxide.

This process is illustrated next:

$$\text{Burn Mg} \rightarrow \text{Heat}$$

$$\text{Burning Mg, in presence of } CO_2 \rightarrow C + O_2$$

$$Mg + C + O_2 \rightarrow MgO + \text{Pure Carbon}$$

Therefore, burning magnesium flares down the smokestack of a fossil-fueled power plant could capture CO_2 and thus preventing it from polluting the atmosphere.

---

[*] https://www.researchgate.net/profile/Andrew_Kindler.
[†] www.green-energy-news.com/content/about.html.

7.9 Oil versus Magnesium: Hunting for Oil and Gas or Looking for Magnesium?

7.9.1 Background

The origin of crude oil goes back to ancient times. It was formed from the remains of animals that died, and trees and plants that were buried and fell to the bottom of the sea. Petroleum oil is believed by most scientists to be the transformed remains of long-dead organisms. The majority of petroleum is thought to come from the fossils of plants and tiny marine organisms.

Oil field operations in general encompass three main phases (as shown in Figure 7.8): oil exploration, drilling operations, and productions; surface petroleum operations; and petroleum refining and fractionation.

The magnesium extraction process, on the other hand, involves the following operations (Figure 7.9):

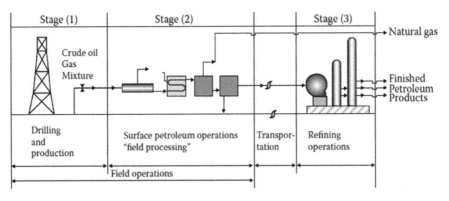

FIGURE 7.8
Oil operations from wellhead to refining.

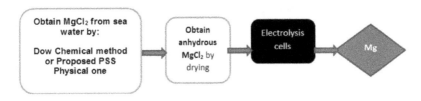

FIGURE 7.9
Magnesium from resources to production.

7.9.2 Analysis and Discussion

Oil exploration is an expensive, high-risk operation. This operation has always been costly and involved risk-taking. The cost of finding oil and gas is increasingly risky and costs are rocketing, which could imperil future supply. The price of each new barrel of oil has risen three-fold over the last decade. The age of cheap oil has gone, and we have to look for alternatives.

With oil, it is a fact that we are facing the dilemma of the rising cost and the search of new oil reserves.

Now is there an alternative to oil?

Magnesium came into the picture rather recently. Looking for magnesium may sound very promising. The fact that there is enough magnesium to meet the world's energy needs for the next 300,000 years is a very good incentive to pursue this line of energy application.

There is no exploration and no risk to find magnesium as compared to oil. Resources have been defined and locations are well established. Magnesium is identified as one of the most abundant elements in the earth's crust. It is ranked number eight. It is the fourth most common element in the earth (after iron, oxygen, and silicon). Surprisingly enough, magnesium makes up to 13% of the planet's mass.

To make the picture clearer, let us put the following facts in black and white:

- Although these figures are highly encouraging, making magnesium an abundant element, its production is neither cheap nor clean.

- Crude oil, once produced as such, has to undergo refining processes to obtain the products that provide us with our need of energy.

- Magnesium, as we have seen earlier, undergoes these expensive steps to obtain an end product. Even so, to make it useful for energy generation, we have to go through additional techniques and elaborate technology: for example: lithium-ion rechargeable cells utilize magnesium as one of the key components in it. Also, when magnesium is connected to make an electric cell to produce electric current, the magnesium becomes the sacrificial anode. Again, if this generated electricity is directed to electrolytic cell, hydrogen will be our source to produce energy.

8

Magnesium: A Potential Hydrogen Storage Medium

8.1 Introduction

By using renewable energies and seawater as the main sources for hydrogen production, very high costs would be incurred for transferring the produced hydrogen from these remote production plants to end-users. Hydrogen can be stored either as a gas or a liquid. Storage of hydrogen as a gas typically requires high-pressure tanks (350–700 bar [5,000–10,000 psi] tank pressure). Storage of hydrogen as a liquid requires cryogenic temperatures because the boiling point of hydrogen at one atmosphere pressure is −252.8°C.

Hydrogen can also be stored on the surfaces of solids (by adsorption) or within solids (by absorption). Compression or liquefaction storage of hydrogen is not practical, especially for intercontinental transport. This approach reduces energy efficiency significantly and can increase hydrogen costs up to a factor of two. Moreover, storing H_2 as a compressed gas or liquid in tanks represents a kind of hazard. Storage of hydrogen in the solid form of magnesium hydrides (MgH_2) in a tank container offers a promising solution in this respect. This would be of use commercially, in order to safely transfer hydrogen from centralized plant sites to intermodal ports. Nanostructured magnesium-based materials have been introduced today in this field. Such materials find applications to form magnesium hydrides. These hydrides represent solid compounds made of hydrogen gas that make some kind of physical–chemical combination with magnesium during the absorption phase. The process is reversible and releases hydrogen when Mg-based material is heated up to a defined temperature. This is the desorption part of the cycle. This allows for hydrogen to be released again into pure gas form. The energy efficiency of this storage system, given by the absorption-desorption cycle, is high, with efficiency around 98.5%.

In a previous theoretical study by the author, he has shown promising results for the economic recovery of magnesium chloride from seawater (PSS), utilizing solar energy in order to store hydrogen by use of magnesium hydrides. Recent developments on solid hydrogen storage methods have opened new, interesting scenarios for an innovative cogeneration process to combine these two ingredients, Mg and H_2. Moreover, recent developments in performance and reliability of water electrolyzers for hydrogen production are creating further positive conditions for such centralized innovative plants to be designed and perform highly sustainable and economically viable H_2 production with subsequent delivering phases whenever it is coupled with sustainable magnesium production.

8.2 Storage of Hydrogen

8.2.1 Background

A brief description is given first for methods of producing hydrogen. Water electrolysis and natural gas reforming are the two technologies of choice. They are proven technologies that can be utilized currently and in the near future. These are recommended to be used as building blocks in the structure of a hydrogen plant for the transport sector. Natural gas reformers of small scale pilot size have been tested in a demonstration project and some are built on a commercial scale-type. However, they are limited in application.

A significant challenge for transportation applications is the building of a hydrogen storage system. This is most needed for both stationary and portable usage. Presently available storage options typically require large-volume systems to store hydrogen in a gaseous form.

Consider automobile manufacturers: They require a lightweight, compact, safe, and cost-effective storage system. On the top of that, it is required to achieve a driving range of at least 300 miles. This driving distance requires a mass of about 5–10 kg of hydrogen as a fuel. The present hydrogen storage technologies available do not satisfy all conditions needed by the auto manufacturers. As a matter of fact, the most challenging problem facing auto makers is finding a solution to the hydrogen storage problem.

Hydrogen is recognized as energy carrier. To produce hydrogen, for example, from natural gas, we have to spend energy to liberate it as a free element, as shown in Figure 8.1.

In addition, hydrogen contains very small amounts of energy on a volume basis. This is because volume is a critical factor when fuel is stored and transported. After producing hydrogen, energy is consumed to store it and transport it to a given destination.

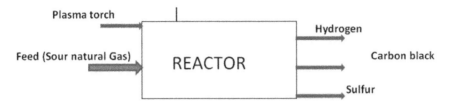

FIGURE 8.1
Schematic presentation of a pyrolysis unit to produce hydrogen.

To summarize, hydrogen can be stored in three ways:

- As a compressed gas in high-pressure tanks
- As a liquid in dewars or tanks (stored at −253°C)
- As a solid by either absorbing or reacting with metals or chemical compounds, or storing in an alternative chemical form

To meet the storage challenge, basic research is needed to identify new materials. Most importantly, problems need to be addressed that deal with the performance of storage systems in question. For example, to store hydrogen as compressed gas, energy must be consumed to reach the high pressure. Compressing hydrogen requires a significant amount of energy, as shown in Figure 8.2.

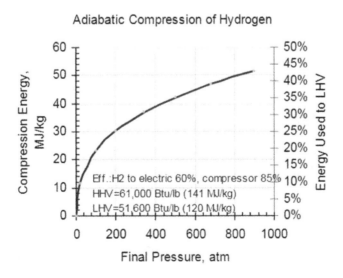

FIGURE 8.2
Energy consumption for hydrogen compression.

This plot shows that compressing hydrogen to 5,000 psi and 10,000 psi requires 36 MJ/kg and 47 MJ/kg, respectively. These correspond to 30% and 40%, respectively, of the low heat value (LHV) of hydrogen. Storing hydrogen as a compressed gas turns out to be non-practical.

The following parameters must be considered for an approach to be pursued for storing hydrogen:

- Operating pressure and temperature
- The life span of the storage material (stability)
- The requirements for hydrogen purity imposed by a fuel cell
- The reversibility of hydrogen uptake and release
- The refueling conditions of rate and time
- The hydrogen delivery pressure
- Overall safety, toxicity, system-efficiency, and cost

8.2.2 Onboard Storage of Hydrogen

On a mass basis, hydrogen has nearly three times the energy content of gasoline. It is 120 MJ/kg for hydrogen versus 44 MJ/kg for gasoline. On a volume basis, however, the situation is reversed; the density of liquid hydrogen to that of gasoline is 8 MJ/L compared to 32 MJ/L, respectively. A comparison of the energy densities for different fuels, based on lower heating values, is given in Figure 8.3.

Onboard hydrogen storage capacities of 5–13 kg will be required as a fuel to meet the driving range for the full range of light-duty vehicle platforms.

To meet the transportation requirements as far as storage is concerned, the storage of hydrogen in chemical compounds offers a reliable method that provides a wider range in this regard. However, no material could provide all the necessary properties needed for application for an AC storage system.

Innovative and basic research is the only thing that leads the way to a breakthrough in materials to be used in storage systems. All scientific disciplines involving chemistry, physics, materials science, and engineering are to be amalgamated to bring a solution for a given storage capacity required. This will include all basic problems to be solved.

At present, there are only three systems for onboard hydrogen storage that could be described as close to commercialization. These are:

- Compressed gas at high pressures (5,000–10,000 psi in composite cylinders)
- Liquid hydrogen, which requires a cryogenic temperature of −253°C

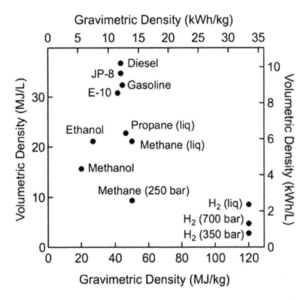

FIGURE 8.3
How fuels compare based on energy content. (From Department of Energy, https://energy.
gov/eere/fuelcells/hydrogen-storage.)

- Material-based storage in solids, which involves the use of metal
  hydrides, carbon-based materials/high surface area sorbents, and/or
  chemical hydrogen storage

The current status of various storage technologies in terms of weight,
volume, and cost is summarized in Table 8.1.

TABLE 8.1

Summary for the Three Modes of Storing Hydrogen

Storage Technologies	Weight (kwh/kg)	Volume (kwh/L)	Cost ($/kwh)
Chemical hydrides	1.6	1.4	$8
Complex metal hydrides	0.8	0.6	$16
Liquid hydrogen	2.0	1.6	$6
10,000-psi gas	1.9	1.3	$16
DOE goals (2015)	3.0	2.7	$2

8.3 Developments in the Storage Medium

For a magnesium-hydride system, the potential as a reversible *storage* medium for hydrogen has led to significant improvements. This basically deals with the hydrogenation and dehydrogenation reaction kinetics. This can be partially achieved by reducing the particle size of the hydrides using ball milling.

The production of a slurry of MgH_2 (which can be pumped in and out, i.e., pumpable), which is safe to handle and releases H_2 by a reaction with water, could be described as an alternative approach. However, it is under investigation. Reprocessing of the $Mg(OH)_2$ into MgH_2 is part of the research conducted for this process.

8.4 The Problems of Using Hydrides as a Source of Fuel for Vehicles

As demonstrated earlier, in order to make hydrogen-powered vehicle's onboard applications to be viable, we need to overcome many problems. Hydrogen-fueled internal-combustion-engine automobiles, today, are getting real. The dream of cleaner, greener transportation is getting real. These vehicles are capable to run hot, finish clean, and produce only pure water as a combustion by-product.

To accomplish this, the industry has to address many infrastructure hurdles that stand in the way of their widespread use. The fuel tank is an example of such hurdles. In search of its technology, hydrogen storage for automobile engines is still undergoing research by many institutions. Greater storage capacities and better gas exchange kinetics than existing models are the targets being thought out by researchers for the next generation of hydrogen fuel tanks.

Using a material like magnesium hydride looks to be a very promising medium for the storage of hydrogen. Simply put, magnesium readily binds hydrogen. This will allow a passenger to use a tank in his car, filled with magnesium, pump in hydrogen, and then use it for his trip. Then, fill it again as needed to run the engine.

Again, the technical hurdle, met by the slow kinetics of adsorption and desorption, when the molecular hydrogen binds to and is released from the magnesium, needs to be resolved.

Research done before has indicated that smaller magnesium nanoparticles have better hydrogen storage properties. This could lead to a solution. Verification of research results has to be done before further implementation is carried out.

8.5 Mechanism of Hydrides Formation

Hydrogen forms metal hydrides with some metals and also alloys, leading to solid state storage under moderate temperature and pressure. The process of hydrogen absorption includes two phases, where in the first phase only some hydrogen is absorbed, and in the second phase, hydride is fully formed.

Properties such as heat resistance, vibration absorption, reversibility, and recyclability are very significant and influence the quality and the costs of magnesium-based hydrides. Moreover, Mg-based materials for hydrogen storage can store a quantity of hydrogen up to 7.7% mass.

The kinetics of hydrogenation of magnesium hydrides are controlled by three factors:

- The rate at which hydrogen molecules dissociate
- Diffusion through the hydride layer formed and into the bulk metal
- Difficulty for hydrogen penetration from the surface into the metal

Dehydrogenation plays an important role in the process. It involves the following consequential steps, or stages:

1. Stages proceeding the bulk, including chemical and structural changes
2. From the bulk to surface transfer
3. Recombination on the surface

It is well established that to enhance kinetics, catalysts can be used. This will improve the process by decreasing activation energy barriers. Nickel and palladium are used as catalysts for hydrogen storage on metal hydrides. It speeds up molecular hydrogen dissociation. Using nickel alloys with palladium is highly recommended. The fast rate of kinetics is obtained because of the catalytic effect of nickel in the alloy.

The composition of some hydrides is given in Table 8.2.

TABLE 8.2

Composition of Some Hydrides

71.5 wt% Mg-23.5 wt% Ni-5 wt% Fe	320	350	3032(abs),2.42(des)
Mg-14 wt%Ni-2 wt%Fe-2 wt% Ti-2 wt% Mo	300	300	4.6
Mg-10 wt%Ni-5 wt%Fe-5 wt% Ti	300	300	5.51(ads) 5.15(des)
$MgH_2 + 10$ wt% TiF_3	300	280	6.27(abs) 5.98(des)
$MgH_2 + 10$ wt% FeF_3	300	280	6.33(abs) 4.82(des)
$MgH_2 - 20$ wt% AB_2 alloy	300	300	5.7
$MgH_2 - 40$ wt% AB_2 alloy	300	300	4.1
Mg-5wt%Ni-2.5wt%Fe-2.5wt%V	300	300	5.67(ads) 4.91(des)
Mg-23.5wt%Ni-2.5wt%Cu	300	300	4
90Mg-6Ni-4C	100	250	5.23(abs) 3.74(des)
$Mg-14Ni-2Fe_2O_3-2Ti-2Fe$	300	300	4.56(ads) 3.32(des)

8.6 Preparation of Magnesium Hydride

According to Wikipedia, preparation of magnesium hydrides from their elements was first reported in 1951. It involved direct hydrogenation of Mg metal at high pressure and temperature (200 atmospheres, 500°C) with an MgI_2 catalyst.

$$Mg + H_2 \rightarrow MgH_2$$

Lower temperature production from Mg and H_2 using nanocrystalline Mg produced in ball mills has been investigated.

Other preparations include:

• Hydrogenation of magnesium anthracene under mild conditions:

$$Mg \text{ (anthracene)} + H_2 \rightarrow MgH_2$$

• The reaction of diethylmagnesium with lithium aluminum hydride

Properties of a *typical* magnesium hydride are reported as follows:

• Compound formula: H_2Mg
• Molecular weight: 26.32

- Appearance: Solid
- Density (Theoretical): 1.45 g/cm^3
- Exact mass: 26.000692
- Monoisotopic mass: 26.0006

8.7 Nanostructured Magnesium Hydrides

The practical use in the development of safe and efficient storage materials for hydrogen represents a challenging issue in this area. Magnesium hydride stands unique in this application. It distinguishes itself among various metal hydrides, owing to its numerous advantages. It could be used as a solid-state hydrogen-storage candidate for application in the hydrogen fuel cell.

In order to upgrade and improve the hydrogen storing properties, it has been proposed to fabricate magnesium hydride *nanostructures* with increased specific surface area and a high number of grain boundaries. The advantages of these changes are:

- Shortening the reaction path
- Accelerating the diffusion rate through the solid phase

Hydriding by using Chemical Vapor Deposition (*HCVD, a Vapor-solid, named, V-S process*) is one of the methods used to produce nanosized magnesium hydride. In this method, the deposition of Mg vapor to form magnesium hydride directly under a pressurized hydrogen atmosphere is applied.

8.8 Preparation of High-Purity Magnesium Hydride by the Second Hydrogenation Method

When Mg powder is hydrogenated under normal conditions, an outer layer of MgH_2 forms first. This MgH_2 layer then prevents hydrogen molecules from penetrating and reacting with metallic Mg, leading to incomplete hydrogenation, even at a high temperature and high hydrogen pressure.

In this article, it was recommended that the outer MgH_2 layer was to be cracked first by mechanical ball milling. Then the milled mixture of Mg and MgH_2 was re-hydrogenated and high-purity MgH_2 was obtained. This method for the preparation of high-purity MgH_2 is called the *second hydrogenation method*. The typical synthesis conditions include a milling time of 9 h.

8.9 Renewable Hydrogen-Magnesium Proposal

Wind and/or solar energy together with seawater can provide all the needed natural sources to produce two basic constituents of the future hydrogen-economy: the green-fuel (namely the hydrogen), and its storage system (the magnesium-based solid hydrides).

The technical viability for producing at industrial scale a large quantity of magnesium and hydrogen is mainly due to a high potentiality in fine-adjusting and integrating current available technologies that are mature in their specific fields. Once hydrogen and magnesium are produced, then the storage of hydrogen using magnesium is one of the main objectives of the RMH proposal. The following are the main two functions to be performed:

1. Water desalination, followed by water electrolysis, to produce hydrogen
2. Magnesium production by the electrolysis of magnesium chloride

The renewable hydrogen-magnesium project (RHM) is, in fact, almost-100% sustainable due to:

1. Total free-carbon primary energy source to fuel the plant units: solar irradiation (or other renewable sources) and seawater only
2. Total reduction of not less than 80% of CO_2 compared to current magnesium extraction via electrolysis processes
3. 100% green-hydrogen production, due to fully green electricity source that powers the water electrolysis unit
4. Solar thermal power concentrated in power generation
5. Water electrolysis in hydrogen production

The RHM project includes the following units/tasks:

- Solar units to generate 100% green-electricity to power the whole process route plant
- Desalination of seawater in order to obtain pure water and salt as by-products

- Water electrolysis to split water in O_2 and H_2
- Magnesium extraction from brine $MgCl_2$ via electrolysis
- Solid-state hydrogen gas storing method by deployment of nano-structured magnesium-based material

The following diagram, Figure 8.4, is an outline for the proposed renewable hydrogen-magnesium (RHM) project.

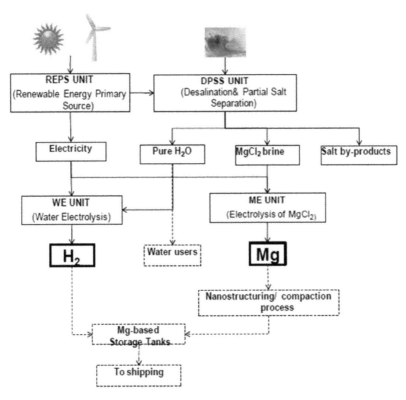

FIGURE 8.4
Renewable hydrogen magnesium proposal.

Appendix A: Magnesium Facts and Uses

Discovery date	1755		
Discovered by	Joseph Black		
Origin of the name	The name is derived from Magnesia, a district of Eastern Thessaly in Greece.		

Fact Box

Fact box			
Group	2	Melting point	650°C, 1202°F, 923 K
Period	3	Boiling point	1090°C, 1994°F, 1363 K
Block	s	Density (g cm^{-3})	1.74
Atomic number	12	Relative atomic mass	24.305
State at 20°C	Solid	Key isotopes	^{24}Mg
Electron configuration	[Ne] 3s^2	CAS number	7439-95-4
ChemSpider ID	4575328	ChemSpider is a free chemical structure database	

Electronic structure of magnesium ion, as shown in Figure A.1.

Appearance

A silvery white metal that ignites easily in air and burns with a bright light.

Uses

Magnesium is one-third less dense than aluminum. It improves the mechanical, fabrication, and welding characteristics of aluminum when used as an alloying agent. These alloys are useful in airplane and car construction.

Magnesium is used in products that benefit from being lightweight, such as car seats, luggage, laptops, cameras, and power tools. It is also added to molten iron and steel to remove sulfur.

As magnesium ignites easily in air and burns with a bright light, it's used in flares, fireworks, and sparklers.

Magnesium sulfate is sometimes used as a mordant for dyes. Magnesium hydroxide is added to plastics to make them fire retardant. Magnesium oxide is used to make heat-resistant bricks for

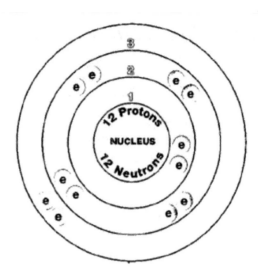

FIGURE A.1
Electronic configuration of magnesium ion.

fireplaces and furnaces. It is also added to cattle feed and fertilizers. Magnesium hydroxide (milk of magnesia), sulfate (Epsom salts), chloride, and citrate are all used in medicine.

Grignard reagents are organic magnesium compounds that are important for the chemical industry.

Biological Role

Magnesium is an essential element in both plant and animal life. Chlorophyll is the chemical that allows plants to capture sunlight, and for photosynthesis to take place. Chlorophyll is a magnesium-centered porphyrin complex. Without magnesium, photosynthesis could not take place, and life as we know it would not exist.

In humans, magnesium is essential to the working of hundreds of enzymes. Humans take in about 250–350 milligrams of magnesium each day. We each store about 20 grams in our bodies, mainly in the bones.

Natural Abundance

Magnesium is the eighth most abundant element in the Earth's crust, but does not occur uncombined in nature. Due to the magnesium ion's high solubility in water, it is the third most abundant element dissolved in seawater.

It is found in large deposits in minerals such as magnesite and dolomite. The sea contains trillions of tons of magnesium, and this is the source of much of the 850,000 tons now produced each year. It is prepared by reducing magnesium oxide with silicon, or by the electrolysis of molten magnesium chloride.

History

Elements and Periodic Table History

The first person to recognize that magnesium was an element was Joseph Black at Edinburgh in 1755. He distinguished magnesia (magnesium oxide, MgO) from lime (calcium oxide, CaO), although both were produced by heating similar kinds of carbonate rocks, magnesite, and limestone, respectively. Another magnesium mineral called meerschaum (magnesium silicate) was reported by Thomas Henry in 1789, who said that it was much used in Turkey to make pipes for smoking tobacco.

An impure form of metallic magnesium was first produced in 1792 by Anton Rupprecht, who heated magnesia with charcoal. A pure, but tiny, amount of the metal was isolated in 1808 by Humphry Davy by the electrolysis of magnesium oxide. However, it was the French scientist Antoine-Alexandre-Brutus Bussy who made a sizeable amount of the metal in 1831 by reacting magnesium chloride with potassium, and he then studied its properties.

Magnesium was used as a curative as early as ancient times, in the form of laxatives and Epsom salts.

In the 1600s, water from the famous Epsom spring discovered in England was a popular curative used as an internal remedy and purifier of the blood. In 1695, magnesium sulfate as a salt was isolated from the Epsom spring water by Nehemia Grew.

Marie de Medici, of the famous and powerful Italian family, described the healing properties of Epsom spring water as, used by *a great store of citizens* especially by *persons of quality*.

Richard Willstatter won the Nobel prize in 1915 for describing the nature of the structure of chlorophyll in plants, noting magnesium as the central element, as shown in Figure A.2.

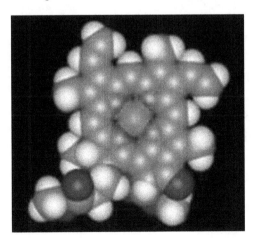

FIGURE A.2
Magnesium is the central element in chlorophyll and the basis of early life on the plant.

Atomic Data

Atomic data								⊘
■ Atomic radius, non-bonded (Å)	1.73			■ Covalent radius (Å)			1.40	
■ Electron affinity (kJ mol⁻¹)	Not stable			■ Electronegativity (Pauling scale)			1.31	

■ Ionisation energies (kJ mol⁻¹)	1st	2nd	3rd	4th	5th	6th	7th	8th
	737.75	1450.683	7732.692	10542.519	13630.48	18019.6	21711.13	25661.24

> Oxidation states and isotopes   ⊘

> Supply risk   ⊘

Others

Unstable in its pure state, magnesium typically forms a white coating of magnesium oxide. In nature, most of its compounds appear as white crystals. Approximately 320,000 tons of magnesium are extracted annually for commercial use.[1,2] Magnesium is commonly extracted from seawater, where it is the third most common component.

Magnesium is the lightest metal with a combination of stiffness and strength.

Pure magnesium has a very high resistance to corrosion because the galvanic activity is low.

Magnesium has great advantages:

- 35% lighter than aluminum, 75% lighter than steel
- Good strength
- High damping capacity
- EFM/RFI shielding (electromagnetic interference and radio frequency interference)
- Good thermal conductivity
- Sustainability
- Stiffness
- Great castability
- Good machinability
- High dent resistance
- Full recyclability

Source

1. Magnesium—Element information, properties and uses. Periodic Table. www.rsc.org/periodic-table/element/12/magnesium
2. Morgo Magnesium LTD is a member of the International Magnesium Association

Appendix B: Common Compounds of Magnesium and Magnesium Ores

Common Compounds of Magnesium

Compound Name	Formula	Molar Mass
Magnesium phosphate	$Mg_3(PO_4)_2$	262.8577
Magnesium sulfate heptahydrate	$MgSO_4·7H_2O$	246.4746
Magnesium chloride hexahydrate	$MgCl_2·6H_2O$	203.3027
Magnesium nitrite	$Mg(NO_2)_2$	116.316
Magnesium ammonium phosphate hexahydrate	$MgNH_4PO_4·6H_2O$	245.4065
Magnesium chlorate	$Mg(ClO_3)_2$	191.2074
Magnesium dihydrogen phosphate	$Mg(H_2PO_4)_2$	218.2795
Magnesium hydrogen carbonate	$Mg(HCO_3)_2$	146.3387
Magnesium cyanide	$Mg(CN)_2$	76.3398
Magnesium hypophosphite	$Mg_3(PO_2)_2$	198.8601
Magnesium sulfate	$MgSO_4$	120.3676
Magnesium hydroxide	$Mg(OH)_2$	58.3197
Magnesium nitrate	$Mg(NO_3)_2$	148.3148
Magnesium chloride	$MgCl_2$	95.211
Magnesium acetate	$Mg(CH_3COO)_2$	142.393
Magnesium hypochlorite	$Mg(ClO)_2$	127.2098
Magnesium phosphide	Mg_3P_2	134.8625
Magnesium hydrogen sulfate	$Mg(HSO_4)_2$	218.4461
Magnesium sulfite	$MgSO_3$	104.3682
Magnesium hydrogen phosphate	$MgHPO_4$	120.2843
Magnesium hydrogen sulfite	$Mg(HSO_3)_2$	186.4473
Magnesium dichromate	$MgCr_2O_7$	240.293
Magnesium iodate	$Mg(IO_3)_2$	374.1103
Magnesium permanganate	$Mg(MnO_4)_2$	262.1763
Magnesium oxide	MgO	40.3044
Magnesium bromate	$Mg(BrO_3)_2$	280.1094

Common Compounds of Magnesium Ores

Natural ores are natural mineral formations that contain magnesium in sufficient quantity to make extraction economically feasible. Magnesium is a component of more than 100 minerals, among them:

Brucite, $Mg(OH)_2$ (41.7% Mg)

Magnesite, $MgCO_3$ (28.8% Mg)

Dolomite, $MgCO_3 \cdot CaCO_3$ (18.2% Mg)

Kieserite, $MgSO_4 \cdot H_2O$ (17.6% Mg)

Bischofite, $MgCl_2 \cdot 6H_2O$ (12.0% Mg)

Langbeinite, $2MgSO_4 \cdot K_2SO_4$ (11.7% Mg)

Epsomite, $MgSO_4 \text{-} 7H_2O$ (9.9% Mg)

Kainite, $MgSO_4 \cdot KCl \cdot 3H_2O$ (9.8% Mg)

Carnallite, $MgCl_2 \cdot KCl \text{-} 6H_2O$ (8.8% Mg)

Astrakanite, $MgSO_4 \cdot Na_2SO_4 \cdot 4H_2O$ (7.3% Mg)

Polyhalite, $MgSO_4 \cdot 2CaSO_4 \text{-} K_2SO_4 \cdot 2H_2O$ (4.2% Mg)

Appendix C: A Letter about PSS

Dear Dr. Abdel-Aal,

It's a pleasure to contact you, as you are one of few referenced people in the world who has competencies on the preferential salt separation process.

I am contacting you as the project coordinator of a co-funded European Project winner of LIFE+2010 Programme. The project is titled "Renewable Hydrogen-Magnesium pilot plant (RHM), and it" officially started on the 1st of January, 2012.

The project concerns the development of a small-scale integrated pilot plant based on an architecture you know very well. The pilot will be powered by a 2000 m^2 photovoltaic roof to feed the three basic units: a desalination unit, to concentrate brine and contemporarily feed pure water to the other electrolyzer unit; the concentrated brine feeds into extract the dehydrated $MgCl_2$, and finally, a magnesium electrolyzer unit, which the dehydrated $MgCl_2$ is fed into extract the pure Mg.

The reason I am contacting you is because the PSS should have been provided by Geo Processors, who was one of our project subcontractors (refer to: http://www.geo-processors.com/technologies/sal-proc.html); I received approval from Dr. Arakel last year to purchase a small SAL-PROC unit.

Unfortunately, before Christmas, Dr. Arakel told me that Geo Processors has been winding down its business and is no longer offering any plant equipment.

As you can imagine, the project is now in crisis; it is a very difficult situation. I received the first advanced payment from the EU Community on the 20th of December, 2011, but the project is now going to stop, and I am worried that it could be terminated by the EU Commission if no substitute for the Geo Processor SAL-PROC unit is found.

Therefore, I would like to ask if you have any useful information regarding other laboratory that could design and develop a small PSS unit, similar to Geo Processors SAL-PROC system?

I would be grateful to you and your partners if you could provide any information about such equipment. In this case, it would be a beneficial for us to have the same contractor designing and developing the two linked units (the Mg electrolysis unit and the Preferential Salt Separation unit).

I am enclosing some information, that I exchanged with Dr. Arakel.

Looking forward to hearing (hopefully good) news from you.

Best regards,
Fabrizio D'Errico
(Project Coordinator LIFE10 ENV IT 323)

Bibliography

Chapter 1

Abdel-Aal HK (1982), Projected economics of a new magnesium production process and their impact on the cost of magnesium hydrides. *Int J Hydrog Energy* 7: 429.

Abdel-Aal HK (2015), Prospects for the role of magnesium in solar-hydrogen energy system. *IJHE* 40: 1408–1413.

Abdel-Aal HK, Al-Naafa MA (1993), Enhanced evaporation of saline water in multi-purpose solar desalination units. *Desalination* 93: 557–562.

Abdel-Aal HK, Ba-Lubaid KM, Al-Harbi DK, Abdullah AS (1990), Recovery of mineral salts and potable water from desalting plant effluents by evaporation. Part I. Evaluation of the physical properties of highly concentrated brines. *Sep Sci Technol* 25: 309–321.

Abdel-Aal HK, Ba-Lubaid KM, Shaikh AA, Al-Harbi DK (1990), Recovery of mineral salts and potable water from desalting plant effluents by evaporation. Part II. Proposed simulation system for salt recovery. *Sep Sci Technol* 25: 437–461.

Chemical composition of seawater; Salinity and the major constituents; energy system. https://www.soest.hawaii.edu/oceanography/courses/OCN623/.../Salinity2015web.pdf (accessed January 24, 2013).

Hussein K, Abdel-Aal HK (2016), *Chemical Engineering Primer with Computer Applications*, CRC Press, Boca Raton, FL.

Kettani MA, Abdel-Aal HK (1973), Production of magnesium chloride from the brines of desalination plants using solar energy, *Proceedings of the Fourth International Symposium on Fresh Water from the Sea*. Heidelberg, Germany.

Zohdy K, Kareem MA, Abdel-Aal H (2013), Separation of magnesium chloride from sea water by preferential salt separation (PSS). *Int J Bioassays* 2(2): 376–378.

Chapter 2

Asian Metal, The World Metal Information Center. www.asianmetal.com (accessed October 6, 2015).

Comstock H (1963), *Magnesium and Magnesium Compounds*, U.S. Bureau of Mines, Washington, DC.

Emley E (1966), *Principles of Magnesium Technology*, Pergamon Press, London, UK, p. 28.

History of Magnesium Production. https://www.nonstopsystems.com/radio/pdf-hell/article-magnesium-prod-history.pdf.

Hussein K, Abdel-Aal HK (2017), *Chemical Engineering Primer with Computer Applications*, CRC Press, Boca Raton, FL.

Magnesium. www.sophisticatededge.com; http://www.chemistryexplained.com/
elements/L-P/Magnesium.html#ixzz4tzZr4f7P.
US Geological Survey (GS) (2014), https://www.usgs.gov/; https://www.usgs.gov/
locations/us-geological-survey-headquarters.

Chapter 3

Balarew C (1993), Solubilities in seawater-type systems: Some technical and environmental friendly applications. *Pure Appl Chem* 65: 213–218.
Bardi U (2010), Extracting minerals from seawater: An energy analysis. *Sustainability*
2: 980–992. doi:10.3390/su2040980.
Craig JR, Vaughtan DJ, Skinner BJ (2001), *Resources of the Earth: Origin, Use,
Environmental Impact*, 3rd ed. Prentice Hall, Upper Saddle River, NJ.
Evaporating seawater and precipitating salts. http://www.hydrochemistry.eu/exm-
pls/sea_evap.html.
http://www.magnesiumsquare.com/index.php?option=com_content&view=arti.
Exploring Our Fluid Earth. Activity: Solubility. https://manoa.hawaii.edu/exp-
lorigourfluidearth/chemical/properties-water/comparison-water-other-liquids/
activity-solubility.
Exploring Our Fluid Earth. Weird science: Types of salts in seawater, https://manoa.
hawaii.edu/exploringourfluidearth/chemical/chemistry-and-seawater/salty-
sea/weird-science-types-salts-seawater (accessed January 2005).
Hussein K, Abdel-Aal HK, Aggour MA, Fahim MA (2015), *Petroleum and Gas Field
Processing*, 2nd ed. CRC Press, New York.
Loganathan P, Naidu G, Vigneswaran S (2017), Mining valuable minerals from seawater: A critical review. *Environ Sci Water Res Technol* 3: 37–53.
Warren J (2015), Salty matters, August 10. www.saltworkconsultants.com/assets/
seawater_evolution-(1-of-2).pdf.
Water Encyclopedia. Mineral resources from the ocean. http://www.waterencyclo-
pedia.com/Mi-Oc/Mineral-Resources-from-theOcean.

Chapter 4

Abdel-Aal HK (2014), From solar hydrogen to desert development: A challenging
approach. *J Chem Eng Process Technol* 5: 6.
Abdel-Aal HK, Abdelkreem M, Zohdy K (2016), Dual-purpose solvay-dow
(Magnesium)Conceptual process. *Open Access Library J (OALIB)* 3: e2998.
Abdel-Aal H, Abdelkreem M, Zohdy K (2017), Seawater bittern a precursor for mag-
nesium chloride separation: Discussion and assessment of case studies. *Int J
Waste Resources* 7: 1. doi:10.4172/2252-5211.1000267.
Wills BA, Napier-Munn T, *Wills' Mineral Processing Technology* (7th ed.). www.
sciencedirect.com/science/book/9780750644501.

Chapter 5

Abdel-Aal (1982), Projected economics of a new magnesium production process and their impact on the cost of magnesium hydrides. *Int J Hydrog Energy* 7: 429.

Abdel-Aal H (2014), From solar hydrogen to desert development: A challenging approach. *J Chem Eng Process Technol* 5: 6.

Abdel-Aal HK (1989), Separation of magnesium chloride from sodium chloride in seawater by the dense-phase technique. *Sep Sci Technol* 24(7–8): 475–484.

Abdel-Aal HK, Ba-Lubaid KM, Al-Harbi DK, Abdullah AS (1990), Recovery of mineral salts and potable water from desalting plant effluents by evaporation. Part I. Evaluation of the physical properties of highly concentrated brines. *Separ Sci Technol* 25: 309–321.

Abdel-Aal HK, Ba-Lubaid KM, Shaikh AA, Al-Harbi DK (1990), Recovery of mineral salts and potable water from desalting plant effluents by evaporation. Part II. Proposed simulation system for salt recovery. *Separ Sci Technol* 25: 437–461.

Abdel-Aal HK, Ibrahim AA, Shalabi MA, Al-Harbi DK (1996), Chemical separation process for highly saline water: 1. Parametric experimental investigation. *I&EC Res* 35: 799–804.

Abdel-Aal HK, Ibrahim AA, Shalabi MA, Al-Harbi DK (1996), Chemical separation process for highly saline water: 2. Systems analysis and modelling. *I&EC Res* 35: 805–810.

Bird RB, Stewart WE, Lightfoot EN (1960), *Transport Phenomena*, John Wiley & Sons, New York.

El-Naas MH (2008), A different approach for carbon capture and storage (CCS). *Res J Chem Environ* 12(2): 3–4.

El-Naas MH, Al-Marzouqi AH, Chaalal O (2010), A combined approach for the management of desalination reject brine and capture of CO_2. *Desalination* 251: 70–74.

Energy requirements of desalination processes, *Encyclopedia of Desalination and Water Resources (DESWARE)*. The contents of this encyclopedia is edited by Darwish M.K. Al-Gobaisi, Editor-in-chief. www.desware.net/Energy-Requirements-Desalination-Processes.aspx. Retrieved June 24, 2013.

Kettani MA, Abdel-Aal HK (1973), Production of magnesium chloride from the brines of desalination plants using solar energy, *Proceedings of the Fourth International Symposium on Fresh Water from the Sea*. Heidelberg, Germany.

Lattemann S, Höppner T (2003), *Seawater Desalination-Impacts of Brine and Chemical Discharge on the Marine Environment*, Balaban Desalination Publications, L'Aquila, Italy.

Mickley M (2004a), *Treatment of Concentrate*. U.S. Department of the Interior, Bureau of Reclamation, Technical Service Center, Water Treatment Engineering and Research Group. (in final preparation).

Mickley M (2004b), *Zero Liquid Discharge and Volume Minimization for Water Utility Applications*. WateReuse Foundation Project. (in progress).

Smith ZA, Taylor KD (2008), *Renewable And Alternative Energy Resources: A Reference Handbook*, ABC-CLIO, Santa Barbara, CA, p. 174.

Chapter 6

Ball CJP (1960), The history of magnesium, *Paper Presented at a Joint Meeting of the Magnesium Association and The Magnesium Industry Council.* London, UK.

Beck A (1943), *The Technology of Magnesium and its Alloys,* Kynock Press, London, UK.

Brooks G et al. (2006), Carbothermic route to magnesium, *J Miner, Metals Mater Soc* 58(5): 51–55.

Celik C et al. (1992), Magnola, an innovative approach for magnesium production, light metals, *Proceedings of CIM.* Montreal, QC.

Comstock H (1963), *Magnesium and Magnesium Compounds,* U.S. Bureau of Mines, Washington, DC.

Derezinski S (2011), *Solid Oxide Membrane (SOM) Electrolysis of Magnesium: Scale-Up Research and Engineering for Light-Weight Vehicles.* MOxST. https://www1.eere. energy.gov/vehiclesandfuels/...2011/lightweight.../lm035_derezins. Retrieved from May 19, 2011.

Eidenson MA (1969), Magnesium, Translated manuscript from Russian.

Emley E (1966), *Principles of Magnesium Technology,* Pergamon Press, London, UK, p. 28.

Faure C, Marchal J (1964), Magnesium by the magnetherm process. *J Metals* 16: 721–723.

Habashi F (1997), 'Magnesium': Handbook of Extractive Metallurgy, Wiley-VCH, Winheim, Germany.

Hansgirg F (1934), Production of metallic magnesium (assigned to American Metals Corp), U.S. Patent 1,943,601, January 16, 1934.

Hoy-Petersen N (1990), From past to future, *Proceedings of International Magnesium Association.* London, UK.

Jenkins DH et al. (2009), Piloting the Australian magnesium process. *Miner Process Extr Metall (Tran Inst. Min Metall. C)* 118(4): 205–213.

JOM, The Member Journal of The Minerals, Metals & Materials Society, https://www.tms.org/JOM.

Magnesium processes for tomorrow—Magnesium.com - Data Bank, www.magnesium.com/w3/data-bank/article.php?mgw=196&magnesium=275.

Magnesium—The Essential Chemical Industry, www.essentialchemicalindustry.org/metals/magnesium.html.

Mayer A (1944), Plant for production of magnesium by the ferrosilicon process. *Trans AIME* 159: 363–376.

Pal UB, Powell AC (2007), The use of solid-oxide-membrane technology for electro-metallurgy. *JOM* 59(5): 44–49. doi:10.1007/s11837-007-0064-x.

Pidgeon LM (1943), Method and apparatus for producing magnesium, (assigned to Dominion Magnesium), U.S. Patent 2,330,142, September 21, 1943.

PNNL- ARPA-E. Extracting magnesium from seawater. https://arpa-e.energy.gov/?q=slick-sheet-project/extracting-magnesium-seawater (accessed February, 2015).

Schoukens AFS et al. (2006), Technological breakthrough of the mintek thermal magnesium process. *J South Afr Ins Min Metall* 106: 25–29.

Simandl GJ, Schultes H, Simandl J, Paradis S. *Magnesium - Raw Materials, Metal Extraction and Economics – Global.* http://www.em.gov.bc.ca/Mining/Geoscience/IndustrialMinerals/Documents/Magnesium.pdf (accessed June, 2016).

Sivilotti OG (1987), Operating performance of the alcan multipolar cell, *Light Metals 1988*, U.S. Patent Nos. 4,514,269 (1985); 4,518,475 (1985); 4,560,449 (1985); 4,594, 177 (1986), The Metallurgical Society of AIME, Warrendale, PA, pp. 817–822.

Solikamsk Magnesium Works, A History, Solikamsk (2001), Annual Report 2015 English web, https://www.google.com.eg/url?sa=t&rct=j&q=&esrc=s&source=web&cd=1&cad=rja&uact=8&ved=0ahUKEwi3yL3mtaraAhWRYVAKHYgYA94QFgglMAA&url=http%3A%2F%2Fxn--glajo.xn--p1ai%2Fraport%2F2016%2F2015_Solikamsk_Magnesium_Works_Annual_Report_web_v.pdf&usg=AOvVaw1cWq9gVdr10Yr4hjtjzh8f.

Wadsley MW (2000), Magnesium metal by the Heggie-Iolaire process. *Mag Technol Miner Metals Mater Soc* 2000: 65–70.

Yan XY, Fray DJ (2010), Molten salt electrolysis for sustainable metals extraction and materials processing–A review. https://www.researchgate.net/publication/51017522_Molten_salt_electrolysis_for_sustainable_metals_extraction_and_materials_processing_a_review (accessed October 19, 2017).

Zang J (2000), The pidgeon process in China and its future, *Paper Presented at the Sinomag Diecasting Seminar*. Beijing. China.

Chapter 7

Abdel-Aal H, El-Shenawy E (2014), Integration of sustainable energy sources with hydrogen vector with case studies. *J Energy Pow Sou* 1(3): 147–151.

Abdel-Aal HK (1982), Projected economics of a new magnesium production process and their impact on the cost of magnesium hydrides. *Int J Hydrog Energy* 7: 429.

Abdel-Aal HK (2015), Prospects for the role of magnesium in solar-hydrogen energy system. *IJHE* 40: 1408–1413.

Dean C (2007), *The Magnesium Miracle*. Ballantine Books, New York.

https://www.crcpress.com/The-Magnesium...Source-of-Energy.../9789814303651 (accessed November 30, 2010).

Magnesium energy cycle system for the power product. https://www.osapublishing.org/abstract.cfm?uri=col-5-S1-S102.

Hussein K, Abdel-Aal HK, Alsahlawi MA (2013), *Petroleum Economics and Engineering* (3rd ed.), CRC Press, London, UK

Kettani MA, Abdel-Aal HK (1973), Production of magnesium chloride from the brines of desalination plants using solar energy, *Proceedings of the Fourth International Symposium on Fresh Water from the Sea*. Heidelberg, Germany.

Magnesium power: White-hot energy—The Economist. www.economist.com/node/15939644.

Magnesium: Alternative Power Source—*Phys.org*. https://phys.org/technology-news/energy-green-tech/.

Magnesium's Massive Power Potential. Aqua Power Systems. aquapowersystems.com/technology/magnesiums-massive-power-potential; http://aquapowersystems.com/aqua-power-systems-update-acquisition-patented-magnesium-air-fuel-cell-technologies/.

Mulliken B (2013), Could magnesium save the world. *Green Energy News* 18(28): 4–5.

Sakurai Y et al. (2007), Magnesium energy cycle system for the power product. *Chin Opt Lett* 5(S1): S102–S104.

Yabe T, Suzuki Y, Satoh Y (2014), *International Conference on Renewable Energies and Power Quality (ICREPQ'14)*, April 8–10. Cordoba, Spain.

Yabe T, Yamaji T (2010), *The Magnesium Civilization: An Alternative New Source of Energy to Oil*, November 30. Pan Stanford, Seattle, WA.

Chapter 8

Abdel-Aal HK (1982), Projected economics of a new magnesium production process and their impact on the cost of magnesium hydrides. *Int J Hydrog Energy* 7(5): 429.

Materials for Renewable Energy & Environment (ICMREE) (2011), *International Conference*. IEEE.

Michalczyk MJ (1992), Synthesis of magnesium hydride by the reaction of phenylsilane and dibutylmagnesium. *Organometallics* 11(6): 2307–2309. doi:10.1021/om00042a055.

Pregger T et al. (2009), Prospects of solar thermal hydrogen production processes. *Int J Hydrog Energy* 34: 4256–4267.

Rajeshwar K (Ed.) et al. (2008), Electrolysis of water. In: H. Kevin (Ed.), *Solar Hydrogen Generation: Toward a Renewable Energy Future*, Springer Science, New York.

Wiberg E, Goeltzer H, Bauer R (1951), Synthese von magnesiumhydrid aus den elementen (Synthesis of magnesium hydride from the elements). *Zeitschrift für Naturforschung B* 6b: 394.

Zaluska A, Zaluski L, Ström–Olsen JO (1999), Nanocrystalline magnesium for hydrogen storage. *J Alloys Compd* 288(1–2): 217–225. doi:10.1016/S0925-8388(99)00073-0.

Index

Note: Page numbers followed by f and t refer to figures and tables respectively.

Milton Keynes UK
Ingram Content Group UK Ltd.
UKHW040051071024
449327UK00019B/474